CARS
That Never Were

BY THE AUTO EDITORS OF CONSUMER GUIDE®

PUBLICATIONS
INTERNATIONAL,
LTD.

Louis Weber, C.E.O.
Publications International, Ltd.
7373 North Cicero Avenue
Lincolnwood, Illinois 60646

Manufactured in U.S.A.

8 7 6 5 4 3 2 1

ISBN 0-7853-0685-4

Library of Congress Card Catalog Number 93-87022

PHOTOGRAPHY
The editors gratefully acknowledge the cooperation of the following people who supplied photography to help make this book possible. They are listed below, along with the page number(s) of their photos:

Special Thanks to: Paul Sichert and Paul Flancbaum, Budd Company; Joe Bortz; Leeland V. Bortmas; David Horton; Auburn-Cord-Duesenberg Museum

Rick Asher, Pontiac Public Relations: 154, 155, 156, 157, 159

Beechcraft Public Relations: 20, 21, 22

Mike Bersch, Brooks Stevens Automotive Museum: 24, 25, 115, 123, 124, 125, 126, 127, 163, 165, 179, 180, 186, 189, 190, 191

Mark Broderick, Chevrolet Public Relations: 57, 63, 65, 67, 68, 69

Elda Burton, American Motors Corporation: 8-19; 191

Robert H. Doehler: 166, 167

Tom Downey, Studebaker National Museum Archives: 148, 165

Helen J. Earley, and James R. Walkinshaw, Oldsmobile History Center: 140, 141, 142, 143

Dan R. Erickson, Ford Photographic Dept.: 98, 99, 100, 101, 102, 103, 104,

Barbara Fronczak and Brandt Rosenbush, Chrysler Historical Collection: 70, 71, 73, 77, 79, 80, 81, 82, 83, 87, 88, 89,

Ford Motor Co. Design Center: 85, 94, 95, 96, 105, 106, 107, 108, 109

Buzz Grisinger: 183

Larry Gustin, Buick Public Relations: 32, 34, 35

Industrie Pininfarina S.P.A.:133

Bud Juneau:172, 173, 175

Floyd Joliet, General Motors Design: 31, 32, 36, 37, 38, 39, 41, 42, 43, 44, 45, 46, 47, 52, 53, 54, 55, 58, 59, 60, 61, 63, 63, 64, 65, 67, 69, 118, 119, 129, 130, 131, 159

Milton Gene Kieft: 96, 97, 132, 133, 150, 151, 152, 153

Vince Manocchi: 177

Doug Mitchel: 19, 75, 127

Mike Mueller: 185

Dick Nesbitt: 111, 112, 113

Chris Purcell, Cadillac Public Relations: 129

Tom Salter: 27, 28, 29

Joseph H. Wherry: 49, 50, 51,167

Nicky Wright: 30, 31, 33, 34, 35, 91, 92, 93, 116, 117, 120, 121, 161, 162, 163, 164, 165, 169, 170, 171

CONTENTS

Detroit has produced countless proposals—like this 1966 AMX II— that never saw the light of day.

Introduction: The Land of Might Have Been

Just as there are many different cars that are or once were, so there are all sorts of cars that never were. And there are more than you might think—certainly more than any one book can thoroughly cover and not wind up as big as a foot locker. So what you'll find here is an interesting sample of cars (and a few trucks) that never made it to your friendly local dealer for one reason or another. In some cases it's just as well they didn't, in others a real pity.

Of course, some never-weres weren't intended for showrooms at all. We're talking about the glamorous one-of-a-kind "dream cars" (or "concept vehicles" nowadays) that most every automaker has used to test public response to new ideas before committing vast sums to produce something similar for sale. Historically, showmobiles have run the gamut from futuristic fantasies, like GM's memorable Motorama exercises of the mid-Fifties, to thinly disguised previews of pending or approved production models, like American Motors' 1966 "Project IV" AMX.

Another class of never-weres involves models that were seriously planned for sale but never got built because a company changed its mind or went out of business. Such stillborns make up the bulk of this book and include everything from ditched Duesenbergs and pitched Packards to forgotten Fieros, a canceled Corvette, a could-have-been *Ford* minivan, assorted "comeback" Studebakers, and an aborted International Scout.

A third category comprises cars produced in such small numbers as to be historical footnotes. They're represented here by the infamous and celebrated 1948 Tucker (just 50 built), the interesting but premature 1959 Charles Townabout electric (production of 200, tops), and Chrysler's bronze-colored mid-Sixties Turbine Car (again 50, though at least those were driven by some 200 people in a "consumer evaluation").

Cars That Never Were also presents a variety of styling ideas for cars that did end up in showrooms with rather different looks. Dating from the Forties, Fifties, and Sixties, these photos record the design evolution of familiar models from Chevy, Ford, and other major nameplates, as well as independents Nash, Hudson, Kaiser-Frazer, and Willys.

Last but not least are the "just for fun" projects pursued with little regard to potential sales or profits—what you might call "studied long-shots." Significantly, most stem from General Motors, which as the world's largest automaker could most afford to indulge in such experiments, at least up to the mid-1980s. And fascinating they are, ranging from tiny electric commuters to a big Cadillac Eldorado "executive wagon."

But to repeat, no one book is big enough to detail every stillborn, speculation, and show car that's ever been, so perhaps someday we'll do a *Cars That Never Were II*. Meanwhile, happy reading and pleasant dreams—or, in certain cases, sighs of relief.

As always, the editors welcome your comments and corrections regarding this volume. Address correspondence to *Cars That Never Were*, c/o Publications International, Ltd., 7373 North Cicero Avenue, Lincolnwood, Illinois USA, 60646.

Chevrolet Corvair Monza SS

AMC's Project IV: Four for the Money

When the going gets tough, automakers get going on show cars. After all, what better than a dazzling "concept vehicle" to dispel any public notions of corporate trouble—and to hold out hope that what's sitting on the spotlit turntable (or something much like it) might soon be sitting in your garage.

American Motors Corporation had plenty of tough going in its 33 years of existence, but for 25 of those years it had the perfect person to head up its styling efforts in Richard A. Teague. A California native and one-time child actor, Teague began his design career in 1948 with General Motors, who hired him after his graduation from the prestigious Art Center College of Design in Pasadena.

But Teague didn't make his mark until after joining Packard in 1951, where he learned to do more work with fewer resources than most any other designer around. Besides the adept facelift for the "last real Packards" of 1955-56, Teague accomplished the memorable Request and Predictor show cars despite the shoe-string budgets then typical of what he later called Packard's "last days in the bunker." The Predictor, of course, was supposed to herald an all-new '57 Packard line that didn't materialize because Studebaker-Packard Corporation was nearly broke.

After brief stints with Chrysler and an independent design firm headed by Ford alumnus Bill Schmidt, Teague went to American Motors in 1959, where he assisted Ed Anderson in what passed for a styling department. When Anderson left in 1964, Teague was named vice-president of design, largely on the strength of his pretty and popular styling for that year's rebodied Rambler American compact. In short order he turned AMC's design section into a far larger and more professional operation.

Meantime, the Sixties were all a-go-go, and AMC was working hard to reverse its stodgy image as a builder of nothing but sensible economy cars. Yet despite the expansionist visions of president Roy Abernethy, to say nothing of Teague's own best efforts, AMC was in financial hot water by 1966. That attracted the interest of one Robert B. Evans, a Detroit industrialist who specialized in corporate rescues and was not, by his own admission, an "automobile man." Still, Evans judged AMC a good investment in spite of its problems, and bought large blocks of company stock. With that, he soon had himself a seat on the board of directors. John Conde, AMC's longtime public relations chief, later described him as "a breath of fresh air at the time."

Evans breezed in believing that the key to AMC's future was "to do things differently—find new ways to do new things and try new ideas." Accordingly, he put Teague to work on what ultimately became a quartet of show cars with plenty of new ideas that promised to jazz up AMC's image in a big way. To ensure plenty of exposure, Evans sent the cars on a nationwide tour as "Project IV," billed as a traveling "auto show of the future."

While none of the Project IV cars saw production per se, one provided a preview of a near-term AMC model. That, of course, was the unique two/four-seat AMX, the direct forerunner of the Javelin-based two-seat fastback that appeared during 1968. The show model had originated in AMC's advanced styling section under Chuck Mashigan in October 1965. Unveiled four months later as a non-running mockup built from a trashed American, it attracted such favorable notice that AMC hired the famed Vignale works in Italy to build a fully operational version for Project IV. It was finished in just 78 days.

Though differing somewhat in details, both the "pushmo-bile" and the Vignale AMXs had the same tight shape—what Teague called a "wet T-shirt look"—plus the whimsically named "Ramble Seat." The latter referred to a pair of jump seats that folded up from the rear cabin floor to provide *al fresco* accommodation for two occasional riders, whose comfort was enhanced by a back window that swiveled up to double as an auxiliary windshield. The Vignale AMX also had a pair of small rear seats inside, for use when the Ramble Seat wasn't. Both show models rode a 98-inch wheelbase like the eventual showroom AMX, and the "runner" carried the same new 290-cubic-inch AMC V-8 that would be standard on production AMXs.

Alas, the Ramble Seat was deemed too costly and impractical for the street. So was another show-car feature: a striking "can-tilevered" front roofline with no visible A-pillars; instead, door glass extended right around to the windshield for an ultra-clean appearance. Somehow, Teague managed to conceal a functional roll bar within. Incidentally, beige leather covered all seats in the runner, which also featured a center console with electric push-button controls for the Ramble Seat.

As we know, the production AMX was a clever section job on AMC's new-for-'68 Javelin ponycar. Even so, it owed much to its one-off forebears in style and character. In fact, when the time came to name it, Teague insisted on keeping AMX even though the initials had originally meant "American Motors Experimental." His logical argument: People already recognized the basic concept by that name, so why change it? But though exciting in itself and downright startling for AMC, the production AMX wasn't all that popular: Just 19,134 were built over three model years.

The other members of Project IV were noticeably less radical than the concept AMXs, yet no less interesting. The least pre-dictive of this forward-looking trio was the AMX II, a smooth notchback hardtop coupe that looked nothing like the Ramble

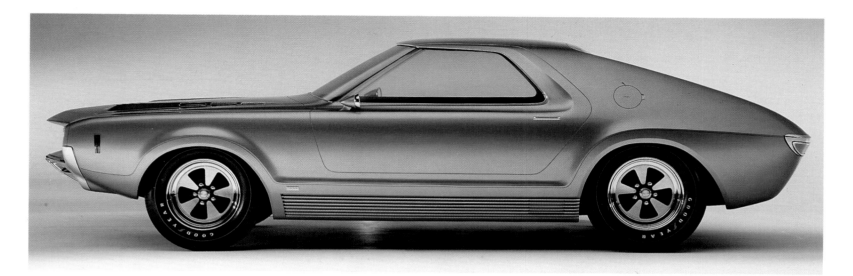

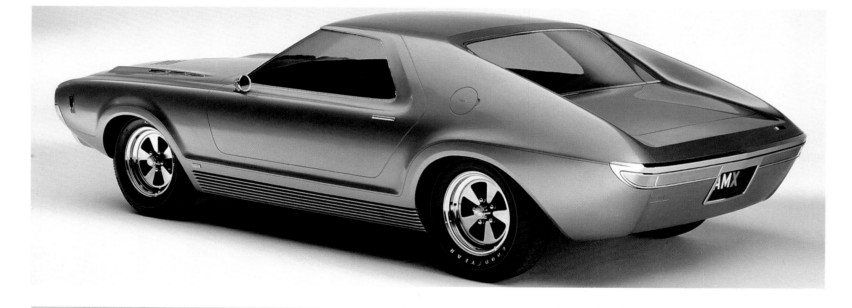

Top: *A profile view of the original AMX "pushmobile." Resemblance to the later Javelin-based production model is clear, but show cars don't have to address practical matters such as realistic bumpers. This car was perhaps the best expression of what designer Dick Teague called the "wet T-shirt" look: smooth, tautly drawn, and compelling to the eye.*

Center: *The AMX "pushbmobile" met such strong favorable response on the Project IV tour that AMC commissioned the famed Vignale works in Italy to build this running version complete with full interior. Front end (left) was simple yet aggressive. Novel "Ramble Seat" (right) provided cozy open-air transport for two. Rear window swung up to double as an auxiliary windscreen. Bottom: Another*

look at the "pushmobile" AMX. Detail differences from the later Vignale-built running model include pull-up instead of pushbutton door handles, silver instead of red taillight lenses—and, of course, no interior. Tunneled backlight and "flying buttress" roof treatment survived in modified form to the 1968 production AMX.

Top: *Though visually related to the Project IV AMX, the AMX II was a full four-seat hardtop coupe created under outside consultant Vince Gardner, not AMC design chief Dick Teague. This profile view emphasizes exceptionally clean detailing and a balanced overall form.* Center: *The AMX II looked good from most any angle, including this rear*

three-quarter aspect. This is actually another view of the green car pictured above, but tinted to give a different color impression, a common trick of car company photo and PR departments. Though not obvious here, AMX II's rear window was slightly vee'd in plan view to match the convex rear-deck line.

Bottom: *By day, the AMX II hid its headlamps behind swing-up doors trimmed to match grille texture. A similar treatment was used on several production models when the AMX II was shown in 1966, including that year's new second-generation Buick Riviera. Big U-shaped bumper was gracefully simple and matched nicely with the pointed front fenders.*

Top: *Four-door Cavalier was the third member of AMC's Project IV—and the most practical. It was built mainly to explore the use of interchangeable parts, which, to Dick Teague's credit, weren't all that obvious. The Cavalier also previewed the general appearance of* AMC's 1970 Hornet compact, particular in its pleasingly simple front- and rear-end styling. Above: *Like other Project IV cars except the Vignale-built AMX, the Cavalier was a non-running model with a "ghost" interior that only suggested seats, dash, and steering wheel.* *Though not seriously considered for production, it was designed to minimize tooling costs. Bumpers, for instance, were the same at each end, as were hood/trunklid panels. Fenders interchanged diagonally; appearances suggested that the doors did too, but AMC never claimed they did.*

Seat cars—doubtless because it was created under Vince Gardner, an outside consultant. Features included hidden headlamps within a big bumper/grille, pointy front fenders, clean flanks, a rear window vee'd in plan view to match deck contour, and "safety taillights" that glowed green, amber, or red to signal going, coasting, or stopping. At 187 inches overall, AMX II was eight inches longer than "AMX I." Most of the extra inches showed up in wheelbase to provide full four-passenger seating.

It's a shame that something like the attractive AMX II didn't wind up as one of AMC's post-1969 Hornet compacts, though Teague did manage a nice Hornet hatchback coupe for '73. He certainly might have remembered it when shaping the swollen '74 Matador coupe, arguably his least distinguished effort after the pudgy-looking late-Seventies Pacer.

On the other hand, Teague did himself proud with the only

sedan in Project IV. Called Cavalier, it looked good for a compact four-door, measuring 175 inches long, with gently curved sides, jaunty "flying buttress" rear-roof quarters, full wheelarches, and minimal "gingerbread." In general form the Cavalier predicted the future Hornet save a shorter hood giving equal-length front and rear decks.

Those proportions were not accidental, for the Cavalier was basically an exercise in interchangeable body panels, an idea that had also lately occupied Brooks Stevens with his "Familia" proposal as a last-gasp salvation for Studebaker. Though less extreme than Stevens's design, Cavalier promised similar savings in production costs. Its hood and trunklid, for example, were completely identical, as were bumpers at each end. Fenders could be swapped left-front to right-rear and right-front to left-rear. Doors apparently did not interchange, though they looked otherwise.

Opposite page: *Vixen was conceived as a 2+2 coupe companion to the Cavalier sedan, but without as much regard for swappable components. Squarish front end (above) gave another hint at the 1970 Hornet, but the applied hood scoop and fastback roofline were unique. Taillight panel (center) is a dead-ringer* for the Hornet's. *"Sugar scoop" rear window/deck was* de rigueur *in the Sixties, and Teague finished it in matte-black for a sportier look. Alas, Vixen's tail seems awkwardly bulky viewed closer to eye level (below)—and woefully "blind," though rear-quarter "gills" concealed small windows.*

This page: *Not part of Project IV, the bold AMX-GT suggests that the later Gremlin would have looked great with a Javelin instead of Hornet front end. Built on a 97-inch wheelbase, the AMX-GT bowed at the New York Auto Show in April 1968 wearing simple flush wheel covers and red paint with a white bodyside/roof stripe (top and center). It was later given five-spoke road wheels and black hood/roof paint (bottom). Rear side glass was fixed, but B-pillars were absent.*

Perhaps just for fun, Teague hinged the rear ones from the back, "suicide-style."

Rounding out Project IV was the Vixen, essentially a coupe companion to Cavalier designed with no thought of panel swapping. It, too, forecast Hornet appearance in its simple blunt "face," dual headlamps, long-hood/short-deck profile, and flared wheel openings. Lower bodysides were a Hornet/Javelin mix, but the distinctive rear roofline was Vixen's own. Small sliding quarter windows enhanced interior ventilation; behind were vertical vanes angled at 45 degrees to give a measure of privacy without obstructing vision. The louvers also conferred a certain design "distinction" that prefigured an unhappy fad of the late Seventies.

Like most auto-show exercises, Project IV was more trial balloon than research tool. Though Abernethy declared that public response "to the innovations presented will have substantial bearing on [our] decisions," many Project IV styling ideas had already been approved for Javelin, AMX, the more distant Hornet, and other future models. Still, evidence suggests that Project IV accomplished its goal of convincing the public that AMC had a future after all, and a bright, exciting one at that.

As a result, AMC experimentals soon became regular showtime fare. For example, 1967 saw an AMX III that previewed Javelin front-end styling as well as the general shape of the Hornet Sportabout wagon then four years away. And though we didn't know it at the time, 1968's AMX-GT was, for all intents and purposes, a preview (warning?) of the 1970 Hornet-based Gremlin.

But AMC's most spectacular experiment was yet to come, a related story that's just a page-turn away.

Opposite page, top: *The 1968 AMX III "sportwagon" show car was also virtually stock '68 Javelin in front. Window treatment differed slightly on the left.* Bottom: *This November 1967 mockup suggests that AMC considered adding a four-door Javelin at some point. A pity the notion went no further.* This page: *AMC experiments from the Seventies.* Top row: *Cute "Concept Electron" was the same electric commuter-car proposal first shown in 1967 as the "Amitron" to announce a joint venture between AMC and battery maker Gulton Industries.* Second row: *Non-running "Concept I" resembled a Gremlin crossed with a Pacer. It, too, was strictly for show. Rear tire bulge was a typical Teague whimsy.* Third row: *"Concept II" was even more Pacer-like—and arguably prettier than what buyers were offered, the smooth front end especially.* Bottom left: *One-off "Grand Touring" was just the Gremlinesque '79 Spirit sedan in more formal dress.* Bottom right: *Neat little "AM Van" was a '79 show attraction that evidently started as a Pacer. AMC might have done well to build copies for sale.*

1970 AMX/3: Nearly A Dream Come True

Dick Teague was always proud that he'd designed the AMX/3—so proud that he snared a third of the cars for himself—which amounted to only two. A lifelong car enthusiast, Teague loved most anything on four wheels, though as a collector he favored vintage machines, which appealed to him as both simple and nostalgic next to the modern iron he worked on during 25 years as design vice-president for American Motors. Even so, those two AMX/3s sat cheek-to-jowl with his big White steamer and massive Pope-Hartford touring right up until his untimely death in 1991.

Designer's vanity? Not at all. The AMX/3 (sometimes written "AMX/III") remains one of the prettiest cars on the globe: low, smooth, curvy in all the right places, adroitly proportioned. Teague had every right to be proud of it.

But good looks are only part of the story. The AMX/3 was also a high-performance mid-engine sports car that came very close to production, which would have been no small achievement even if volume would have been scarcely more than the six examples ultimately built. Still, while Chevy teased the public with midships Corvettes that would never be, little AMC was briefly on the verge of building a Euro-style supercar the public could actually buy.

The story begins in 1969 with the AMX/2. Like the Big Three, American Motors was then aggressively courting the youth market, and the Teague studios had issued a variety of "think young" showmobiles in pursuit of same, as well as groovy new showroom offerings like the Mustang-inspired Javelin, the unique two-seat AMX, and the outlandish SC/Rambler. AMX/2 was something else. Patterned after European exotics like the Lamborghini Miura, Lotus Europa, and Porsche 914, it was not just AMC's most daring showmobile ever but one of Detroit's first acknowledgements that mid-engine design was The Coming Thing in production sports cars.

That the AMX/2 got built at all stemmed from the enthusiasm of AMC group vice-president Gerald C. Meyers and chairman Roy Chapin, Jr. The shape came from a Teague sketch that had taken Meyers's fancy: a two-passenger fastback with what the designer termed an "airfoil" shape. The eventual non-running fiberglass mockup sported a "fast" windshield, shapely downcurving nose with functional hood vents and hidden headlamps, and a raised rear-deck "spine." The last provided pivot points for twin tiltable spoilers; its outboard ends were flared neatly into the rear fenders. Though Teague supervised these and other details, the design was executed by staffers Bob Nixon and Fred Hudson.

Though clearly just a pipe dream, AMX/2 was greeted with no little interest on its public unveiling at the Chicago Auto Show in February 1969. With people promising to buy if only AMC would oblige, Meyers and Chapin decided to take the next logical step by commissioning a fully engineered version that could be built for sale in at least limited numbers. But just to make sure they weren't missing something design-wise, they also decided to solicit a proposal from Italy's Giorgetto Giugiaro, then increasingly regarded as the world's most talented car designer.

Giugiaro's team duly produced a full-size mockup in lightweight foamcore that was shipped to Detroit for "comparison competition." Never publicly shown, it bore the low, angular lines then typical of Giugiaro, but looked lumpy next to the Teague group's model, which resoundingly carried the day.

Yet despite its all-AMC design and engineering, the resulting AMX/3 was quite European in many ways. For example, noted race-car engineer and builder Giotto Bizzarrini supervised chassis development in Italy, and BMW assisted with testing. The four-speed transaxle in the first AMX/3 came from ZF in Germany, though OTO Melara of La Spezia, Italy was later tapped for a new gearbox and final drive that could better withstand the hefty torque of the installed AMC 390 V-8.

Per mid-engine practice of the day, that V-8 mounted longitudinally behind a snug two-seat cockpit, with the transmission trailing behind. Suspension was by classic all-around double wishbones and coil-over-shock units, with dual springs at the back and an anti-roll bar front and rear. Brakes were big four-wheel vented discs from Germany's Ate. Also like many midships contemporaries, the AMX/3 used different-sized front/rear rolling stock: 205-15 Michelin X radials on 15x6½ Campagnolo alloy wheels fore, 225-15 tires on massive 9-inch-wide rims aft. Dimensions were quite compact: 105.3-inch wheelbase, 175.6-inch overall length, 74.9-inch width. Tracks were fairly generous at 60.6/61.2 inches front/rear. Overall height was just 43.5 inches, yet ground clearance was a respectable 5.9 inches.

Despite weighing some 3100 pounds, the AMX/3 had an estimated top speed of 160 mph, thanks to its 340-horsepower engine and fairly short 3.45:1 final drive. Unfortunately, high-speed stability was none too good. Teague later recalled that Bizzarrini drove an AMX/3 on the demanding Nurburging in Germany and "became nearly airborne at 145, so that kind of slowed him down. . . . It did get very front-end light."

Even so, the AMX/3 almost made it to AMC dealers. First shown to reporters in Rome on March 23, 1970, it would have replaced the Javelin-based AMX, according to Teague, but "in a much more contemporary vein and not [sharing] anything with the Javelin. . . . And the price would have been $10,000 instead of $4000, [so] it would have been more of a prestige car, kind of an image-building car. We were into racing at that time with Trans-Am and all that, and it was really kind of a tool, but a serious one,

16

Unveiled at the Chicago Auto Show in early 1969, the non-running AMX/2 surprised everyone as the first evidence that tiny AMC was toying with an advanced mid-engine sports car. Styling, created under company design chief Dick Teague, featured an "airfoil" shape, and managed to be interesting—swoopy without being cartoonish. Graceful simplicity was the watchword from curved nose to "Kamm" tail. A fastback roofline with a dorsal "spine" floated above a flat, ribbed engine cover-cum-deck. Central twin exhausts implied a V-8, which was planned for but not installed in this purely speculative exercise. Still, public response was wildly enthusiastic.

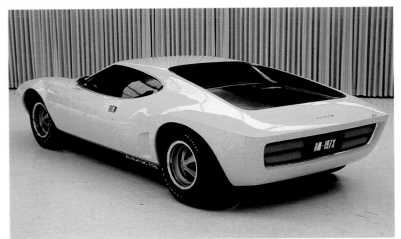

to create an image for the company that was something other than four-door Ramblers and 'Ma and Pa Kettle' cars." The specific plan was to build 24 units in 1970, then increase output gradually in line with demand. But AMC's continuing sales problems, projected engineering costs for meeting new federal safety standards, and difficulty in securing a body supplier all conspired to put AMX/3 on the shelf after just six examples were built, all effectively pre-production prototypes.

Happily, all six survive today. Number-one wound up in the Gilmore Museum in Kalamazoo, Michigan, numbers two and four in the Indianapolis area. Teague owned numbers three and five, but their present whereabouts are unknown (most of his collector cars were sold after his death). The final AMX/3 was completed sometime later (likely during 1971) at the behest of a business friend of Bizzarrini's. As you'd expect, there were slight detail differences among these cars, but number-six was the only

AMX/3 with concealed wipers and three extra inches of rear overhang. That Italian businessman saw to the cutting-up of two unused bodies, which later prompted Teague to speculate that a couple more examples might surface someday.

That, alas, seems unlikely, but at least we have the six AMX/3s to inspire thoughts of this grand midships *turismo* that might have been. For us, they will also ever inspire memories of the good friend we lost in Dick Teague, one of the great unsung talents of the design business. It says much about this man that he could always laugh at mistakes like the late-Seventies Pacer as graciously as he accepted praise for brilliant successes.

The AMX/3 was definitely Dick Teague's kind of modern automobile, and his enthusiasm showed in every line. As unquestionably his finest work, it's the one we should remember him by as both car designer and enthusiastic "car nut." He would want it that way. He deserves no less.

Opposite page: *Evolved partly in response to high interest in AMX/2, the AMX/3 was fully engineered for limited production in 1970-71 as an image-building $10,000 replacement for the Javelin-based AMX. Dick Teague's initial design ideas* (center left) *survived to final form with virtually no alteration* (top row, *as seen in AMC's design showroom), though wheels and other details differed among the six cars ultimately built. Journalists went to Rome in March 1970 to get their first drives* (center and bottom right). *All were impressed. So, too, Mark Donohue* (bottom left), *then winning Trans-Am races in factory Javelins. This page, top: Taken during the Rome press preview, this publicity photo shows what appears to be the number-three car later owned by designer Teague. Bottom: His other AMX/3 was number-five, shown here in the early Eighties, still pristine in its original bright yellow paint.*

1946 Beechcraft Plainsman: No Wings, No Prayer

Cars are obviously quite different from aircraft, yet corporate experience with one can encourage a flirtation with the other. Henry Ford, for example, put America on wheels with his beloved Model T, made a fortune, and went on to produce the all-metal Tri-Motor passenger plane. Henry's venture into flight was no mere dalliance, for his "Tin Goose" gained a stature equal to that of the Model T, and helped launched regular U.S. airline service in the Twenties. Abroad, BMW was famous for aero engines long before it built cars. So, too, was Bristol of England, which didn't turn to automaking until the late Forties—and with prewar BMWs at that. Other firms like Daimler-Benz, Hispano-Suiza, and even Rolls-Royce were involved simultaneously with cars and aircraft at various times. So there was plenty of precedent for the Beechcraft Plainsman, "an aircraft manufacturer's idea of what an automobile should be."

Development of the Plainsman began in late 1945, not long after V-J Day. America was still savoring final victory over the Axis, and American consumers—denied new cars for nearly four years—were ready to buy. The result was a frenzied seller's market that attracted all manner of would-be auto tycoons, who reasoned that all they needed to succeed were a little money and lots of *chutzpah*.

The Plainsman surfaced in 1946, looking like one of those wildly predictive "cars of the future" that were staple elements of wartime and postwar tech magazines. The car was novel, all right, but Beech Aircraft Corporation had a more critical motive for eyeing the automobile business: survival. Like other aviation companies in 1945-46, Beech faced life with no new military contracts at hand, and dim prospects for civilian aviation sales. Accordingly, the firm spent $50,000 to devise a "super-modern" passenger car that might conceivably save the corporate neck should aero sales stay grounded. It might even teach Detroit a thing or two.

What emerged was a large, heavy-looking four-door fastback sedan that resembled a Czech Tatra with the cockpit of a Japanese Zero. Dominating the form was a tall aircraft-type greenhouse emphasizing upward visibility via auxiliary panes above the windshield and door windows. This was separated from the main glass by thin bars that appeared to wrap around from one back door to the other. The windshield itself was divided but radically curved both outboard and above. Doors were cut into the roof to ease entry/exit, foreshadowing both the Tucker and the far more distant '63 Corvette Sting Ray coupe. The high roofline made for towering head room despite chair-high seats, while the rangy external dimensions gave plenty of maneuvering room for the shoulders, elbows, and legs of six adults. The bench-type driver's seat offered four-way electric adjustment, still a novelty in 1946. To the right of that seat was a separate two-person "double lounge."

Even more than spaciousness, the Plainsman interior stressed safety and simplicity. Door panels, for example, were flat and protrusion-free. (Solenoid-activated pushbuttons substituted for conventional door handles both inside and out.) Instruments and controls were sensibly grouped around the steering wheel in an otherwise spare-looking dash, though Beech planned to include an "economy meter" giving a continuous mpg reading, and even a two-way mobile telephone (hence the circular roof-mount antenna). The speedometer was in direct driver view above the steering column in a small, black, hooded pod. There were no seatbelts—surprising for an aircraft maker—but leather-covered rubber "crash pads" adorned the dash and the tops of the front seats. The roof structure was allegedly quite strong despite its slim A-pillars, which also aided visibility. Looking straight back or over-the-shoulder was tricky, however, given the very wide, "blind" roof quarters and a relatively small rear window.

Only one Plainsman prototype was built. Though it wore a conventional grille, it carried a rear-mounted engine: an air-cooled, horizontally opposed gasoline four adapted from one of Beech's contemporary aircraft units. But the compact powerplant could have been up front just as well. Not only was there plenty of room for it, but Beech planned on using an innovative four-wheel electric drive system that completely eliminated the differential, propshaft, clutch, and transmission (and with the last, the interior floor hump, thus adding to passenger room). Exact details of this patented system were never disclosed—likely because it *was* patented—but it's known to have worked from a generator driven by the engine and housed in the same soundproof compartment. Wheel control was evidently independent, because Beech claimed two advantages for its electric drive: automatic apportioning of torque to those wheels with greater grip on slippery roads—call it embryonic traction control—and a "reverse current" feature that provided "dynamic braking" when the driver stepped on the pedal—a sort of early anti-lock system. Naturally, there were also regular hydraulic drum brakes, activated by a slightly harder push on the same pedal.

Only a bit less novel was the Plainsman's air-spring suspension. Unlike later Detroit "air ride" setups that relied on flexible bladders, this one employed aircraft-type air shocks that automatically adjusted damping to suit load and weight distribution. A manual override switch was planned to permit selecting a softer setting for a smoother ride on very rough roads. Though not known for sure, the damping rate was probably varied by the drive system's electric generator.

Coming from an aircraft maker, the Plainsman was predictably designed for low weight and good aerodynamics in the interests of both performance and economy. Aluminum was chosen for exterior panels and inner structure, and the shape was test-

Beech likely chose the Plainsman name to emphasize the links between its aircraft and this prototype car. Top: Front end bore a conventional grille even though the prototype used a rear-engine format, but the car was designed so a front engine could be used just as easily if desired. Above left: Another period publicity photo emphasizes the Plainsman's airplane-style "cockpit" window treatment and spacious six-passenger interior. Large window areas and thin A-pillars were trumpeted as safety pluses for visibility. Above right: High roofline apart, the Plainsman was just another late-Forties "bathtub," though wind-tunnel tests confirmed the aerodynamic efficiency of its rounded lines. Despite a small engine, the prototype was claimed capable of over 160 mph and up to 30 mpg.

either mode, 68 mph combined. Acceleration was just as tepid: 0-40 mph, for instance, took well over a minute on the little gas engine alone, 36.4 seconds with the torquier electric motor (which could produce up to 20 horsepower in brief spurts), and 21.9 seconds in hybrid mode. "At that rate," said Sherman, "merging onto a busy expressway would be heart-stopping."

But forget freeways. The B&S Hybrid was conceived as a surface-street runabout that would cost peanuts to run and hardly smoke up the atmosphere. Though emissions from the gas engine were apparently neither controlled nor measured, economy was outstanding. B&S claimed 25-52 mpg in gas-only mode, and said that the recommended use of the two powerplants—electric to accelerate, gas engine for cruising—was calculated to give up to 85 mpg. However, as Sherman noted, the "pure-electric range" was only 30-60 miles and a deep recharge took 6-8 hours.

Then again, this was only an experiment in high economy born of fears that gasoline would become far more scarce and costly—if not today, then tomorrow. It was also something of a trial in how little raw power a future car might be able to get away with and still be practical. Not that Briggs & Stratton ever intended to enter the car business, though it would have doubtless gladly welcomed a Ford or General Motors building hybrids in the millions—and paying $643 apiece for little air-cooled B&S twins to power them. Of course, the electric motor was sold separately and batteries were not included.

But none of this matters now that hybrid power has been all but excluded from our future by legislative decree. With battery and electric-motor technology advancing again by the Nineties, the State of California decided to "drive the market," as it had with emissions controls in the Sixties, by mandating that automakers sell a certain percentage of "zero-emissions" vehicles by the year 2000 as a condition for doing any business within its borders. Other states seem sure to follow this lead, so research now focuses almost exclusively on pure-electric vehicles—and let the dirty old internal-combustion engine rest in peace.

Which is really too bad. Though *Car and Driver*'s Don Sherman derisively likened the B&S Hybrid to "what you get when you mate a garden tractor with a golf cart," he felt its concept held genuine promise. As he wrote in late 1980, it had the makings of a new Citroën 2CV "[with] the same . . . B&S engine, front-wheel drive, two axles, Kip Stevens bodywork, a carbon-fiber chassis and a 1200-pound curb weight. It should sell for $3000 (or less) [contemporary dollars] and deliver 50 mpg. The Ace Hardware store could sell a bunch of 'em."

Then again, maybe not. Remember the Crosley or the Allstate? Case closed.

Above left: *The Briggs & Stratton prototype employed a hybrid drive comprising an air-cooled gas-fired two-cylinder engine, made by B&S, and an equally small Baldor electric motor. Both lived up front along with ancillary* mechanicals. Top right: *Hybrid's unusual six-wheel chassis was borrowed from an existing small electric van. Rearmost wheels helped support almost a half-ton of batteries mounted above, just below a shallow luggage hold. Only the* middle axle was driven. Above right: *Shown with an early-1900s electric car, its spiritual forebear and part inspiration, the B&S Hybrid survives today in the Brooks Stevens Museum just outside Milwaukee, where this photo was taken.*

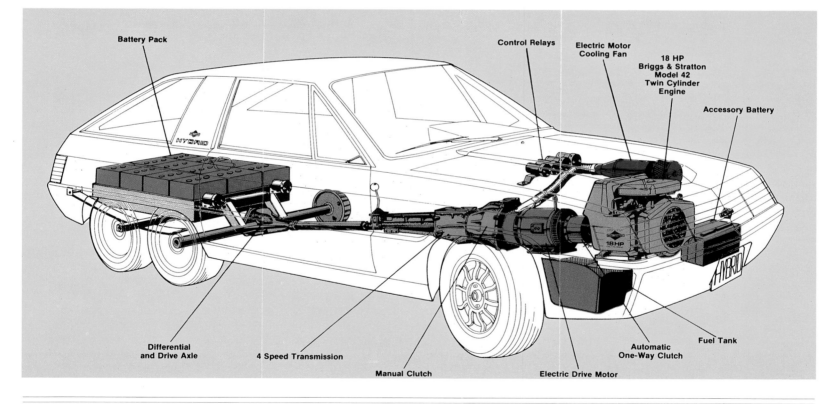

Battery Pack

Control Relays

Electric Motor Cooling Fan

18 HP Briggs & Stratton Model 42 Twin Cylinder Engine

Accessory Battery

Differential and Drive Axle

4 Speed Transmission

Manual Clutch

Electric Drive Motor

Automatic One-Way Clutch

Fuel Tank

Top: *The Hybrid in another period photo taken on the grounds of the Stevens Museum. Young Kip Stevens, son of noted industrial designer Brooks, handled the styling, which managed to look good despite the awkward six-wheel chassis. Body was rendered in fiberglass to help offset the heavy battery pack for best mileage and performance. Overall appearance vaguely recalls VW's Scirocco. Note the covered headlamps.* Above: *The Hybrid bares its novel drivetrain in this illustration of major components. A Borg-Warner "Duo-Cam" clutch allowed separate or tandem use of the two power sources. Drive in all cases went via a conventional four-speed manual transmission borrowed from a Ford Pinto. Note the massive bank of 12 car-type batteries over the third axle.*

close to the XT-Bird estimate. So was promised delivery: about six months, which Budd was pleased to note was the time taken to design, engineer, and build the prototype.

Budd's arguments for the XR-400 also echoed its XT-Bird pitch—all "pro," of course: a recent uptick in foreign sports-car sales, mounting demand for "personalized" bucket-seat compacts like Chevy's Corvair Monza, steady growth in the number of two-car households where "the second car is frequently of the sport type," the widespread appeal of such cars . . . and, yes, continuing high demand for used two-seat T-Birds. Still, Budd allowed that most Americans were put off by foreign sports cars because of high cost, limited passenger room, weak resale values, and scarcer parts and service; it also acknowledged a public "preference for American-made products." But these only further argued for XR-400: "After all, it would be made in this country . . . servicing and parts would present no problems [with AMC's broader dealer network] . . . the price would be right . . . and [XR-400] would provide passenger space available in no other comparable car." That last statement was true on its face, but even Budd's own photos showed the rear seat was too small for anyone but toddlers.

No matter. Budd lobbied as hard for the XR-400 as it had for XT-Bird. Besides engaging in some unabashed client flattery, it sketched a scenario in which AMC would pioneer a market "presently untapped by any other manufacturer," with a car so "unlike anything else on the road it would attract widespread attention . . . provide your dealers with both a new profit area and morale-builder . . . and offer unusual advertising and sales promotion opportunities."

All these arguments were much like the ones Iacocca was using to muster the Mustang. And why not? The evidence was there for everyone to see. Yet for all the kind words and rosy pic-

ture-painting, AMC was unmoved. So like XT-Bird, XR-400 was a no-sale.

AMC's decision reflected its dwindling profits of the day and the arrival of new president Roy Abernethy, who had his own ideas about cars, sporty and otherwise. Judging by his practical dictates for the unfortunate "3+3" Marlin of 1965, Abernethy doubtless deemed the XR-400 too small to sell well against other sporty compacts. Then too, AMC had just started selling bucket-seat cars of its own, and designer Dick Teague had sculpted a pretty new Rambler American convertible for 1964, a full four-seater that appealed more to Abernethy as a likely profit-maker.

In the end, AMC was probably right to ignore the XR-400. Even if buyers had accepted its rather sedate Budd styling, the limited interior package would have been a tough sell. Besides, a sports car, however "practical," was the last thing people expected from AMC—or that AMC dealers knew how to push.

But tantalizing questions remain. Would a Rambler sports convertible have been as hugely popular and profitable as the Mustang? Would AMC be with us now had it introduced the XR-400 some six months before Mustang arrived—say in late 1963, as Budd said was possible? We'll never know.

We do know that the XR-400 made two public outings soon after AMC refused it. Renamed "XR-Budd," it helped tout the firm's then-new disc brake at the Detroit Auto Show and Society of Automotive Engineers annual convention in early 1964.

The XR-400 resides at the Budd Company today in good original condition despite the passage of three decades and some 13,000 miles of sporadic driving. The firm has also preserved the XT-Bird. Let's be glad both these "might-have-beens" are still around as reminders of how easily history can be made when you just say no.

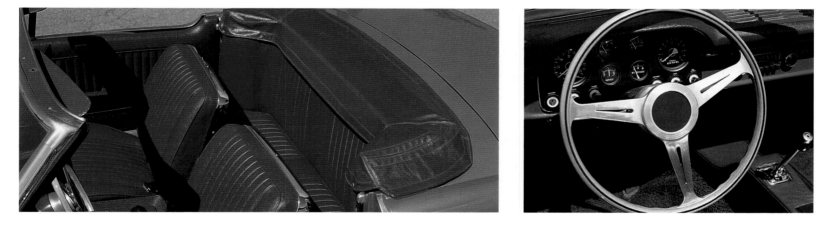

Opposite page: XR-400's least lovely aspect was probably its "face," but AMC might have fixed that and other matters had production gone ahead. The prototype now wears chrome-reverse rims but originally had simple full wheel covers. This page, top row: Proportions made for a rather ungainly profile, but the long wheelbase and short overhangs were good for interior space. Top-up appearance suggested a two-seater, but the layout was actually 2+2. Center: Simple dash was not unlike the first Mustang's Falcon-derived panel. Semi-bucket front seats came from AMC, as did many other hardware items. Above left: Though cramped except for tots, the XR-400's rear bench seat was likely included because Ford had turned down Budd's earlier XT-Bird for lacking four-place seating. Above right: XR-400 put minor controls and full gauges dead-ahead of a three-spoke wood-rim steering wheel in classic sports car fashion. Ignition switch on the extreme left followed then-current Ford fashion.

Buick's Sporting Show Cars: Trying Two-Seaters On for Size

Though not originally called a Buick, the experimental Y-Job of 1938 was built on a contemporary Buick chassis modified by division chief engineer Charles Chayne. Its Harley Earl styling forecast the look of Flint's all-new 1942 models. Still rightly called the "granddaddy of all dream cars," the Y-Job survives today at the Henry Ford Museum on a permanent loan from GM.

Top left: *The Y-Job has seen many changes over the years, such as this 1949-style Buick dash, which was fitted about that time and remains today.* Top right: *Y-Job displays its predictive hidden headlamps in a rare hood-up vintage photo. The lights were later exposed, and remained so at least through the early* *Seventies before the car reverted to the original "face" it still shows.* Center left: *The Y-Job circa 1940 with Harley Earl, GM's legendary first design chief, at the wheel. Low stance was achieved partly with 13-inch road wheels, uncommonly small for 1938.* Center right: *Another period picture shows the Y-Job's rakish* *"boattail" trunklid.* Above: *Y-Job has been seen over the years with and without fender skirts; it wears them today. A metal cover always cleaned up top-down appearance, and would be featured on future Earl show cars. It's since become a fixture of many production convertibles.*

Top: *Buick had two dream machines in 1951: the finny, aircraft-inspired LeSabre and, pictured here, the more conventional XP-300. Bodies made heavy use of aluminum and magnesium, and both cars carried an experimental aluminum V-8 that was supercharged to make 335 horsepower. Unlike LeSabre, though, XP-* 300 *forecast future showroom Buicks in its frontal styling, which was echoed on the '54 models.* Center: *Buick chief engineer Charles Chayne takes the XP-300 for a spin. And why not? He was one of its "fathers." Wheelbase was a trim 115 inches. Note the wrapped windshield and ribbed side trim.* Above left and right: *The one-off Wildcat bowed at the 1953 Motorama in convertible form, but GM designers gave it a lift-off hardtop by early '54, as shown in this rare factory photo from mid-January. "Roadmaster" appeared on the base of the top just above the rear deck.*

Top: *After long years of obscure neglect, the 1953 Wildcat was rescued and fully restored by Chicago-area collector Joseph Bortz. A fiberglass body rides atop a stock '53 Roadmaster chassis cut to a 114-inch wheelbase. Unusual "Roto-Static" scooped hubs remained fixed as the wheels turned to feed cooling air to the brakes. Frontal styling previewed '55* showroom Buicks. Above left: *The Wildcat's aft view is vintage Harley Earl: jazzy, with a hint of jet fighter. Note the prominent trunklid fins and protruding exhaust tips. Above, upper right: Green leather bathes the Wildcat's two-seat cockpit. Dash design hints at Buicks to come. Above, lower right: The Wildcat helped showcase Buick's new* 1953 ohv "Fireball" V-8 with 322 cubic inches and 188 horses. The one installed in the Wildcat got lots of chrome to catch the show lights, and required little work during restoration. Unlike many showmobiles that are fragile and incomplete, the Wildcat had many standard parts and was well built. Bortz says it drives like any period Buick.*

Top: *Buick followed up the '53 Wildcat with the logically named Wildcat II of 1954, a smaller two-seater in the mold of Chevy's lately introduced Corvette sports car. Wheelbase was 100 inches, two inches less than the 'Vette's; overall height was a rakishly low 48.5 inches. Happily, Wildcat II also survives today,* and this is how it looks. Above left and right: *This is how it looked in a 1954 GM publicity shot. Genuine knock-off wire wheels and dashing cutaway front fenders were "classic" counterpoints to a trendy wrapped windshield and the liberal use of chrome typical of the age. Look hard at the profile and you might* think this was just a customized Corvette, but it wasn't. Rear end echoed the look of Buick's full-size limited-production '54 Skylark convertible, though the chrome "banana" trim would also show up in modified form on the 1958 Corvette.

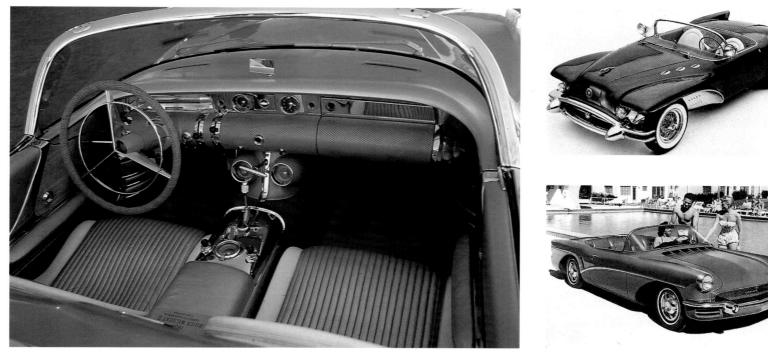

Top: *Conceived largely by veteran Buick designer Ned Nickles, the Wildcat II originally had swiveling cowl-mounted "spot" headlamps astride its windshield, but they've since been moved to a more conventional spot beneath the cutaway front fenders. The original "greyhound" hood mascot has evidently been lost along* the way. *Otherwise, the car is still much as it was in 1954. Above left:* Like "Wildcat I," the Wildcat II was *somewhat predictive of near-term Buick dash design, but center console was unique. Above, upper right: Another look at Wildcat II in 1954. Bearing Buick's trademark "Ventiports," the hood* concealed a 322 V-8 with four carbs and 220 bhp. *Above, lower right: The 1955 Wildcat III heralded some elements of Buick's '57 production styling in a trim four-seat package on a 110-inch wheelbase. Flint doubtless flinched when Ford scored big with its very similar four-seat Thunderbird in 1958.*

BUICK'S DISCARDED FLIGHTS OF

FANCY FROM THE FIFTIES,

RECENTLY UNEARTHED IN THE

GENERAL MOTORS ARCHIVES,

ARE FASCINATING.

Buick's Basement: Designs We Didn't See in the Fifties

Buick's 1954 styling was well underway as early as 1951. Oval headlamp nacelles from that year's XP-300 show car appear on a scale model from January '52 (top left), while a full-size mockup from

December '51 (top right) shows early conceptions of the rounded rear fender treatment that was also ultimately adopted. Wraparound windshield was never in doubt. Designers evidently held

out a long time for heavy side trim (above left) and thick grille bars (above right), as on the 1952-53 models, but a cleaner look with squarer contours finally prevailed.

Top three rows: *Buick's all-new '57 body was supposed to prevail for the usual three-year styling cycle, but when Chrysler stole the design limelight with its flashy, finned "Forward Look" '57s, GM had second thoughts about 1959. Its intent through January '57 had been to offer facelifted '58s that year. These studio "record shots" show some '59 Buick ideas under that plan. Most* involved even busier trim on the already garish "B-58" models, plus big outrigger front-bumper bombs that made for a rather "bottom-heavy" look. It's just as well GM changed its mind, as these cars would likely have fared no better—and maybe worse—than the slow-selling '58s. *Above, left and right: The "crash" redesign for 1959 decreed greater* divisional body-sharing to save both time and money. Initial work was underway by February '57, when Buick designers modeled these full-size clays. Greenhouses are close to final form, but the eventual "delta wing" fins had yet to surface. Front fender profile on the hardtop coupe is much like that of the production '59 Cadillac.

Top two photos: *Among the more startling finds in the GM vaults is this close-coupled hardtop coupe badged "Skylark II." Photographed in the Design Staff courtyard on August 19, 1957, it bears most hallmarks of the eventual '59 Buicks save a heavier-handed rear end.*

Though this might have been a belated reply to Ford's two-seat Thunderbird, it was surely forgotten when the four-seat '58 Bird scored far higher sales. Remaining photos: *A group of side-elevation tape drawings from October '57 with suggested trim variations for the '59*

Buicks. Basic shape is close to final, but note the blunted fins on two of these and the Mercury-like rear roof on the four-door sedan, as well as the use of traditional Buick series names, which evidently weren't abandoned until fairly late in the game.

Top two photos: *By April 1958, GM stylists were working on their 1960 models, which were inevitably planned as facelifts of the all-new "crash-designed" '59s. Contemplated for Buick were heavily browed, widely spaced headlamps* and a rather Chevy-like rear with flattened fins and broad horizontal taillamps. **Center and above:** *Three months later, Buick staffers tried a slightly different tail, resurrected "Ventiports," and moved the headlamps* closer within a new front resembling that of the 1960 Mercury. The '60 Buick emerged as more conservative yet bulkier-looking than any of these proposals, with fins rounded off and heavy side sculpturing.

Great Expectations: Cadillac's Postwar V-12 and V-16

Cars with more than eight cylinders have always been scarce. Since World War II, Detroit has produced the V-10 Dodge Viper, but the others have been costly foreign exotics. But though America has yet to field a modern V-12 or V-16, Cadillac at least entertained the notion back in the mid-Sixties.

And why not? After all, Cadillac's experience with such cars dated from distant 1930, when its V-16 bowed at least a year ahead of every other domestic multi-cylinder engine. The cars it powered have long ranked among the greatest of the Classics, and the V-16 itself was arguably the most important engine to emerge from that era save the hallowed Duesenberg J straight-eight. What's more, only Cadillac really profited with multi-cylinder power. Before demand tailed off in mid-1930, General Motors's prestige division shipped over 2000 Sixteens, more than all the 16-cylinder Marmons built in three years. Lincoln and Packard didn't have anything comparable until 1932, and then "only" V-12s. By that point, Cadillac had a V-12 of its own, and the make's combined multi-cylinder sales were always the industry's highest by far, if modest by absolute standards.

But multi-cylinder giants were an arrogant indulgence for "hard times," and most were gone well before World War II. The main reasons were high production costs and selling prices versus very low demand, which forced Packard and even Cadillac to abandon such cars entirely after 1940. Though Lincoln's small "cheap" V-12 persisted through 1948, it was obsolete long before.

An added blow to multi-cylinder American cars in the postwar period was the advent of high-octane gasoline, which allowed smaller engines to produce comparable power via overhead valves and higher compression, thus eliminating the need for more than eight cylinders. With that, the ohv V-8 became Detroit's engine of choice. By 1955, every U.S. make offered at least one.

Nevertheless, the expanding prosperity and sky's-the-limit optimism of the Eisenhower years soon renewed thoughts of 12- and 16-cylinder engines, only with modern, high-compression ohv heads. Leading the way was Packard, which in 1955 was desperately seeking to recover some of its squandered past glory. The idea was a V-12 derived from its just-announced V-8. According to former product planner Richard Stout, this would have been machined on the V-8 line, the longer block being moved "halfway down" to bore the extra cylinders. Since the V-8 block was a 90-degree "Y," 30 degrees out of phase for the "in-step" firing desirable in a V-12, each rod throw would be staggered 30 degrees to compensate. Buick would use this same "split-throw" principle for its 90-degree V-6 in the early Sixties.

With 480 cubic inches and all the horsepower that implied, Packard's postwar V-12 would have been a mighty work indeed. But as Dick Stout remembered: "It was strictly grandstand stuff. . . Tooling was guesstimated in the $750,000 area—modest for such a spectacular result. . . . But in the end the money just wasn't there." In fact, Studebaker-Packard then faced imminent bankruptcy and so abandoned luxury Packards after 1956, substituting medium-priced Studebaker-based cars through the marque's sad demise in 1958.

A few years later, Cadillac Division took its own stab at modern multi-cylinder power. The attempts followed two paths: a fairly crude, "bolted together" V-16 composed of two V-8s, and an exotic all-new V-12 with single overhead camshaft.

The contemplated V-16 probably had nothing to do with Cadillac's own V-8. By 1960, the division's milestone 331 V-8 of 1949 had swelled to nearly 400 cid, which would have made a twin-block sixteen simply gargantuan. A more likely choice was Chevrolet's 283, which would have doubled-up to 566 cid—big, but not impossible. According to former GM Design Director Chuck Jordan, who then headed Cadillac Styling and worked on that side of the V-16 revival, this engine was more conjecture than concrete proposal. "We were working with Engineering Staff to put two V-8s together," he later recalled. "It was kind of a homemade way to do it, but it was just to project an image we wanted to get across to [division management] at the time. Nothing serious was ever developed engineering-wise."

More intriguing was the clean-slate V-12 being prepared at the same time. Jordan remembered this as "a very sophisticated powerplant, and quite beautiful. I'm sure it was designed from scratch as an overhead-cam engine—a very exciting piece of machinery to see." Also a 90-degree unit, it was designed for in-step firing a la Packard's stillborn V-12. Aside from that and sohc heads, technical details remain obscure.

It's equally unclear whether either of these engines was ever seriously considered for production, but there's no doubt that development stopped at the prototype stage. Still, Jordan and his colleagues came up with a remarkable group of design studies for a new multi-cylinder Cadillac. One of their first scale models was a close-coupled coupe with approved 1963 lower-body styling but a much longer front—truly enormous, in fact—plus a dramatically tapered fastback roof of the sort favored by corporate design chief William L. Mitchell. Gradually, work proceeded through scale- and full-size clay models designed from the ground up.

Jordan emphasizes that it was all mostly for fun: "It was a two-passenger Cad with 16 cylinders all right, but it was done just to make a statement about the heritage of Cadillac and where we were going with the image. . . . We built several scale models and one full-size clay. The concept of all was invariably the same: a long-hooded car to contain the long engine. These designs were exaggerated, almost cartoon-like, but exciting to work on. This was one of our pet advanced projects at the time."

Top left and right: *Early work toward a new multi-cylinder Cadillac produced this scale model, photographed at the GM Technical Center to look more like the real thing. Probably fashioned around 1961, it wears by-then approved '63 lower-body styling, but also a vastly longer front to accommodate the engine, plus a sweeping fastback roof. The last would likely be too much for Cadillac's* conservative audience even now. Second from top: *From May 1963, a crisp yet bulky-looking V-12 hardtop proposal with close-coupled notchback roofline and wild two-tier front. Aft fenderlines would show up in modified form on Caddy's new front-drive '67 Eldorado. Second from bottom: Rather reminiscent of Classic Thirties speedsters, this circa 1963 concept for an open V-16 wears* "Cyclone" badges. Note the plethora of hood air intakes. Rakish perimeter windscreen harmonizes with Buick-style bodyside "sweepspear" lines, but would have been far too costly for series production. Above: *A more formal multi-cylinder concept by designer Wayne Cady from March 1965. Semi-open fastback would have been a Sixties novelty. C-post script intriguingly reads* "LaSalle."

Still, somebody must have taken this work seriously, for by December 1965 that full-size V-16 clay was known as one of GM's famous "Xperimental Projects:" XP-840. As photos testify, designers really cut loose on this two-passenger fastback, giving it a massive undercut nose, semi-separate front fenders, a huge wrapped windshield *sans* A-pillars, double-notched beltline, a rear "view port" with TV camera instead of a conventional window, and a back panel deeply inset between "outrigger" fenders.

Scale models were no less wild, yet some contained ideas that would later show up in production. One fastback combined lower body contours like those of the 1966 Olds Toronado with an exaggerated version of the '66 Buick Riviera front. Renderings often toyed with open or semi-open concepts. Wayne Cady, for example, sketched a sleek targa-style fastback with skirted rear fenders, rakish windshield, and a definite Thirties flavor. Another drawing shows a super-low roadster with the character of a speedboat: low windscreen wrapped all the way to the rear fenders, vee'd deck, 16 "ventiports" in the long hood. The last, presumably, were air intakes for separate carbs in these years before government emis-

sions standards and Detroit's wholesale move to fuel injection.

In the end, though, there was no hope. As Jordan recalled: "We finally dropped the project after the full-size model was completed. We had a lot of other things to do, and here we were playing with a full-size clay we never intended to expose. It was strictly a styling exercise."

Of course, it's styling exercises like this that fire the dreams of enthusiasts, and in retrospect it's a shame the division went no further with a postwar multi-cylinder car. Yet with V-12s now available from Jaguar, BMW, and Mercedes, to say nothing of Lamborghini and Ferrari, could a V-12 or V-16 still be in Cadillac's future? Hard to say, but such a car would surely enhance the make's stature in a way the late two-seat Allanté never could seem to manage. On the other hand, newer Cadillacs like the Northstar-powered Seville STS have accomplished a good deal of image restoration, so perhaps the division will never again venture beyond V-8s.

But we hope we're wrong. A new multi-cylinder Cadillac could be a thing of wonder—a new "standard of the world."

Another scale model for the stillborn new-generation multi-cylinder Caddy, this time from August 1963, but again artfully photographed to look "real." Prominently peaked hood and grille are evident in all three shots here. Overhead perspective (above right) highlights a "cantilevered" roof sans A-posts;

radically vee'd windshield; a vee'd rear deck to complement both windshield curvature and the pointy snout; plus muscular, wide-stance proportions. Yet how much longer and slimmer it seems at "ground level." Profile (top) displays hints of the 1966 Buick Riviera front, while the forward three-quarter view

(above left) displays sharp fenderlines and bulging wheel arches like those adopted for Oldsmobile's new '66 Toronado. Fastback roof treatment with no beltline "break" at the rear wheels would also be seen on both those production models, though in quieter form.

By December 1965, the new multi-cylinder Cadillac had won "official" status as project XP-840 and progressed to this full-size mockup for a two-seat V-16 fastback coupe. Typically shown in the GM Design viewing court, it continued the "substantial" look of earlier proposals, most notably in profile (top), plus a domed "prow" hood and

matching vee'd windshield (above). But there were many startling firsts here—like no back window or inside rearview mirror; instead, a narrow slit was cut into the roof as a viewport for a rear-facing TV camera (center left and right). Note also the bold "outrigger" rear fenderlines, semi-separate front

fenders, and the ribbed "cuffs" spilling out and down from the hood to recall the outside exhaust pipes of Classic days. Nameplates here read "Eldorado," but probably just for convenience. Cadillac had no need for a new V-12 or V-16, and all the design work toward such a car was done mainly in speculative fun.

In Search of Eldorados: The '71 "Shooting Brake"

GM evolved this handsome Eldorado wagon during 1969 after the '71 hardtop coupe and reborn Eldo convertible were locked up. Exterior appearance was handled by Cadillac's Wayne Cady working under Clare MacKichan, the GM veteran who then headed the firm's Advanced Design Studios. More speculation than serious production prospect, the project went only as far as a single full-scale clay model, shown here in initial finished form, plus a fully working interior seating "buck" created under Ed Donaldson, who remembers that the entire studio staff spent a lot of time on the wagon.

Top: *More than for any utilitarian hauler, Ed Donaldson's interior concept for the '71 Eldo wagon combined Cadillac luxury with unusual versatility. As shown here, the front bucket-seat backrests were intended to fold flat to* form *a makeshift bed with the rear buckets, whose backrests could be flopped forward to extend the load platform. Numerous amenities were housed along the quarter panels. Rear package shelf was fixed.* Above: *The '71 Eldo's* transformation *to semi-sporting "shooting brake" was a remarkable and rather surprising aesthetic success. Vee'd back panel continued the rear-window theme of first-generation 1967-70 Eldo coupes.*

THE ELECTRIC CAR MAKES

PARTICULAR SENSE TODAY. YET

ELECTRICS NEVER REALLY WENT

AWAY, AS THIS INTRIGUING LATE-

FIFTIES EXPERIMENT ATTESTS.

1959 Charles Townabout: California Dreamin'

It's taken the better part of 70 years, but electric cars are again big news—and available. At least they are and will be in air-polluted California, where lawmakers have told automakers to sell 40,000 "zero-emissions" vehicles starting in 1998 and some 200,000 by 2003. It's a rather draconian edict evidently inspired by the movie *Field of Dreams:* If you tell them to build electric cars, they will, never mind the cost, technical problems, and other unresolved matters.

Of course, Southern California has ever been a hotbed of automotive invention as well as the smog capital of the world, so it's no real surprise that various vocal locals tried sparking public interest in "volts wagons" long before legislators did. San Diego, for example, was home to one very practical electric car way back in 1959. Just as interesting, the project originated in the region's second best-known industry after filmmaking, the then-booming aircraft field. Even its name had the quaintly memorable quality of a movieland character: Charles Townabout.

The first part of that name honored Dr. Charles Graves, executive vice-president at Stinson Aircraft Tool & Engineering Corporation, the postwar descendant of the famed Stinson Aircraft Company. But he was not your average corporate bigwig. Besides being a dentist, Graves had credentials as both physicist and electronics engineer. As such, he took due note of recent predictions that "super" storage batteries were just around the corner, followed soon perhaps by fuel cells that could make electricity by means other than combustion. Graves was also aware of electric-car experiments going on at places like the Cleveland Vehicle Company in Ohio and, closer to home, the Nic-L-Silver Battery Company up in Santa Ana, California. With the right technology and a little luck, he reasoned, Stinson might just beat everyone else to what loomed as a huge and lucrative new market for electric cars that could carry the firm through the ups-and-downs of its aircraft business. Accordingly, Graves put Dean Van Noy in charge of engineering an amps-powered auto, assisted by Dick Bardsley and Graves himself.

The result appeared by June 1959 looking like a mildly customized Volkswagen Karmann-Ghia coupe—which essentially it was. There was even a similar torsion-bar suspension. Still, the Townabout had plenty of differences. Most naturally involved motivation, which comprised four 12-volt car-type batteries linked in series to twin electric motors that were coupled directly to the K-G's rear halfshafts—one motor to drive each wheel. Gears were cut for a 6:1 rpm reduction and beveled for quiet operation. The motors, supplied by Baldor of St. Louis, developed 3.2 shaft horsepower apiece, roughly equal to 11 bhp from an internal-combustion engine. To keep weight down, the VW's steel "backbone"

frame was replaced by a special box-section aluminum chassis with a welded-on platform to serve as both floorpan and body carrier. Curb weight ended up at about 1800 pounds. The light weight, sprightly gearing, and 22 total horsepower produced acceleration along the lines of popular period economy cars—like the Karmann-Ghia. Naturally, there was no power lost to a transmission or differential because they weren't needed and were thus omitted. Also missing was the K-G's token back seat/parcel shelf, the space being used instead to stow the bulky battery pack. The motors and allied components lived farther behind, in what had been the VW's engine bay.

Veteran auto journalist Joe Wherry later wrote that, appearances notwithstanding, the Townabout body was made entirely of fiberglass molded from a stock Karmann-Ghia. However, bulkheads, floorpan, roof, windshield pillars, and the battery deck were formed as a single unit for strength, matching that of the lightweight chassis. To keep the Charles from being confused with Wolfsburg's car, Graves and company remodeled the nose a bit and etched in a below-the-belt "character line" that took a saucy dip before kicking up into trendy tailfins. Townabout prototypes wore a fake grille and rather contrived bumpers. Later versions were spared the former, got small Ford-like round taillamps in the fins, and carried simpler bumpers comprising two wraparound chrome tubes connected by a third tube curved into an inverted U.

As one of the few journalists then covering the electric-car scene, Wherry got to drive a pre-production Townabout and came away impressed. Stinson's craftsmanship was high quality all the way, and particularly good on the fiberglass work. Performance was at least adequate. Though the Townabout was gear-limited to a top speed of only 55 mph, Wherry said tests in San Francisco showed hill-climbing ability equal to that of large-displacement piston-engine cars.

Wherry also liked driving electric, finding it easier than a conventional car. All you did was turn a key, select "Forward" or "Reverse" motor operation, and step on the "gas." No shift or clutch to bother with, no noise to speak of—and no engine braking, though just letting off the "throttle" slowed things down quite fast, as anyone who's driven an electric golf cart can attest. Inside were a pair of sporty bucket seats and just three instruments: speedometer, the all-important ammeter, and a "battery condition" gauge showing the amount of charge remaining. Stinson also fitted conventional windshield wipers and turn signals but no radio, which was "not recommended" for the Townabout or most other budding electrics. Perhaps the firm had nightmares of people leaving their radios on too long and hope-

The Charles Townabout "volts wagon" in mid-1959 as snapped by the late Joe Wherry, one of the few auto writers then covering the electric-car scene. High-quality body was molded in fiberglass from a contemporary VW Karmann-Ghia coupe, hence the obvious visual resemblance. This is one of the final pre-production cars that Wherry drove, marked by a clean nose instead of the fake grille tried at first. Created by Stinson Aircraft in San Diego, the Townabout saw production of only about 200 units, plus a dozen or so prototypes. Intended volume was much higher, but Stinson backed out due to limited potential sales and a necessarily steep retail price of $2895.

ONLY A HANDFUL OF DESIGN

IDEAS CAN BE USED FOR ANY NEW

MODEL. CHEVY HAD PLENTY OF

INTERESTING ONES IN THE

FORTIES, FIFTIES, AND SIXTIES.

The Way They Weren't: Styling Ideas Chevy Threw Away

Like other makes, Chevy was planning its mid-Forties cars when Pearl Harbor thrust America into World War II. This undated full-size mockup suggests what a 1943 or '44 Chevy (note the hubcaps) would have looked like had the war not come along. Though the simple bar grille is similar to what was actually used for 1946 (above right), the profile (top) and rear views (above left) show an evolution of the flow-through fenderline first seen on certain 1942 Buicks. Per longtime company practice, this would have been extended to other models within a few years, Chevy included, but the war postponed it until GM's first all-postwar generation of 1948-49. Note the broad lower body moldings and the neat "shadow box" license-plate recess in the rear bumper.

Top left: Looking a bit Dodge-like, this hardtop coupe was sketched for Chevy's all-new '55 line, but with a shorter roof and longer deck than finally chosen. This may have been the "Bel Air Executive Coupe" idea that inspired the '58 Impala. Modest tailfins were later planed off. **Top right:** *A retractable Chevy hardtop? This January '55 model sure looks like it, with a lower windshield and longer, higher "bustle" rear deck. If GM had been tipped about Ford's '57 Skyliner, it didn't follow through.* **Second row from top:** *Early 1954 saw these trim workouts for Chevy's '56 facelift. Note that the rear bumpers wrapped around and very far ahead.* **Bottom two rows:** *This full-size clay from early 1955 displays facelift ideas for '57. Modeled as a two-door on the left side and a four-door on the right, it wears front-fender "windsplits" and an embryonic "bullet" front bumper, but contributed little else to final '57 styling. Still in the running at this point were holdover '56 taillamps and the '55-style eggcrate grille favored by GM design chief Harley Earl.*

Top and center: *Three tape drawings from October 1957 show the '59 Chevy nearing final form. This was roughly 10 months after GM scrapped original plans for '58 facelifts in favor of all-new styling prompted by Chrysler's successful 1957 "Forward Look." Graceful profile and* wild "bat-fin" rear deck were settled on quickly and not changed much later, but Chevy stylists long toyed with Edsel-like fronts bearing centrally stacked high-beam lamps. Mercifully, that motif was discarded for a more orthodox "face" that looked miles better. Above: *A trio of* rear-end workups for the full-size '64 Chevy, dated February 1962. These early thoughts clearly aimed at a more impressive version of '63 styling, which was already locked up (note the model in the corner of the studio in the photo on the right).

More GM Design "record shots" show how the big 1964 Chevy was evolving a bit later in '62, with efforts still focused on the popular Impala Sport Sedan. Above right: First thoughts involved a light facelift of approved '63 styling (seen in photo at rear of studio) with the same crisp fenderlines. "Face" here has the same general appearance as that worn by Chevy's new '64 mid-size Chevelle. Other photos: The big '64's lower body soon took on a huskier, more rounded look, and the grille went from oblong to "dumbbell" shape to highlight the headlights. Triple-taillight motif was already an Impala hallmark and thus never in doubt. Round-light panel was ultimately retained (as in top left), but "browed" vertical treatment (above left) would have been interesting.

Sweet Dreams: Those Memorable Corvair Specials

The decline and death of Chevrolet's rear-engine Corvair is one of the sadder ironies in automotive history. Though this lively yet economical compact would be perfect even in today's world, it was disparaged by consumer advocates—the very people who might have been expected to welcome it.

But Ralph Nader was not responsible for killing the Corvair. The ax had actually fallen some six months before the release of his 1965 book, *Unsafe at Any Speed.* General Motors's decision to continue the car beyond '65, but only until development and tooling costs were amortized, reflected the success of Ford's Mustang "ponycar," which overwhelmed Corvair in the sporty-car market the way Ford's Falcon had swamped it in the economy field. It was Mustang and not Nader that forced GM to rethink its ideas about bucket-seat sportsters, which eventually led to the Chevy Camaro as its proper Mustang-fighter. Of course, Corvair was more technically advanced than either Mustang or Camaro, which partly explains why it still inspires enthusiasm.

The Corvair also inspired designers and engineers to use it as the starting point for several fascinating experiments. Some came surprisingly close to reaching showrooms.

Super Spyder, 1962

GM's first Corvair special was the 1961 Sebring Spyder, a jazzy, short-wheelbase preview of the '62 Corvair convertible. Though basically stock below the beltline, it boasted low racing windscreens and fixed door windows, a two-seat cockpit with "backbone" divider bar, and long dual headrests faired into the rear deck. A year later came the Super Spyder, a wilder evolution of this basic design on the same 93-inch wheelbase (15 inches less than stock).

Like its predecessor, the Super Spyder wore a tonneau behind its cockpit, but with a single driver's headrest in a tapering pod *a la* Jaguar's famed late-Fifties D-type and SS sports/racing cars. A trio of vertical louvers rode ahead of each rear-wheel opening as dummy air scoops (the Sebring had a pair of functional slots in that spot). Triple chrome exhaust pipes exited from behind each rear wheel, which prompted thoughts of the late Hudson Italia. The drivetrain comprised a 150-bhp turbocharged six and four-speed manual transmission from the production Monza Spyder.

All told, the Super Spyder was a good-looker, more dramatic than the Sebring, and the most advanced Corvair special built up to that point. But GM never even hinted that it might be built.

Monza GT and SS, 1962-63

GM designer L. W. Johnson got a look at Bertone's two-seat, Corvair-based Testudo as it was nearing completion in 1962-63, and happily informed the Italians about his own company's latest Corvair special. But though built in 1962 and tested informally that year at Elkhart Lake and Watkins Glen, the new Monza GT fastback wasn't publicly unveiled until April 1963, when it handily stole the limelight at the New York Auto Show along with a racy open sibling called Monza SS.

Both these experiments were fiberglass-bodied two-seaters with four-wheel disc brakes, magnesium-alloy wheels, hydraulic clutches, stock four-speed-manual Corvair transaxles, fixed seats, and adjustable pedals. Both were also cleanly styled in the billowy idiom then favored by GM, with bumperless fronts and oblong headlights concealed behind large "clamshell" lids.

Yet for all their similarities, the GT and SS differed markedly in some ways. The GT coupe, for instance, carried its engine ahead of the rear axleline; the SS roadster put it behind, as in production Corvairs. Wheelbase was 88 inches on the SS, 20 inches shorter than stock, but 92 inches on the more visually aerodynamic GT. Like production Corvairs, the roadster offered a small front luggage compartment, but the coupe did not. Inside, the SS dash was stark but highly informative, with a large tach and speedometer, plus five auxiliary gauges. Had a production version materialized, the low, racing windshield and fixed side windows would have been replaced by a conventional screen and roll-up glass.

The SS also had normal doors where the GT used a Testudo-style lift-up cockpit canopy, again front-hinged at the cowl and extending back to the B-pillar region. Also echoing Bertone's work was the rear-hinged hatch that swung up to reveal the engine, which was a two-carburetor version for quiet, smooth running. The SS used a four-carburetor setup.

Though often credited solely to GM design chief Bill Mitchell, Monza GT and SS styling was actually the work of Larry Shinoda, now celebrated for his work on various Corvettes and early-Seventies Mustangs, and Anatole "Tony" Lapine, who would go on to design Porsche's 928. Both the GT and SS were created under project code XP-797. The SS at least came fairly close to actual production—as close as any Corvair special. Remember that when these Monzas first appeared, Ford hadn't released the Mustang and Corvair was still GM's only low-cost sporty car. Had the redesigned '65 Corvair sold better against the Mustang, it's not inconceivable that a street SS or GT could have appeared by 1967 or '68—which means GM might not have needed to create the Camaro, with all the expense that entailed. But the Mustang sold like nickel hamburgers from day one, and GM quit working on future Corvairs entirely.

Which was a shame, because the Mustang wasn't half the sports car these Monzas might have been. As *Road & Track* said at the time: "The enthusiast market sorely needs a boost, and these are two cars that could do it." Come to that, they still could.

Top left and upper right: *Though it bore a hint of Corvette Sting Ray, the 1963 Corvair-based Monza SS roadster looked like nothing else: smooth, well-formed, and ultra-clean. Shape and detailing were vintage period GM, yet also somewhat Italian. Styling was executed by Larry Shinoda and Tony* Lapine under GM design chief Bill Mitchell. *Upper right: The Monza SS with its coupe companion, the Monza GT. Both mounted fiberglass bodies on modified Corvair chassis and were strictly experiments, though the SS was briefly eyed for production until Ralph Nader came along and Corvair sales* plummeted. *Above: Unwrapped in 1962, the Super Spyder was a more radical version of 1961's Corvair-based Sebring Spyder. Race-car design cues are obvious, and a stock 150-horsepower turbocharged Corvair six delivered performance to match.*

Vetting Corvettes: Stillborn Styling for America's Sports Car

Top left and right: *The Corvette caused quite a stir on its 1953 debut, but not many sales through '55, when GM bean-counters nearly killed it. Just before that, company design chief Harley Earl, who'd helped fashion Chevy's sports car and remained one of its biggest boosters, suggested this mild facelift for '55 or '56, crafted on a production '54 model. The main changes involved a neat eggcrate* grille, front-fender vents, and a sculpted "shadow box" rear deck as on the 'Vette-based '54 Motorama Corvair coupe. Ultimately, GM managers decided to try harder for '56, greatly encouraged by the stunning all-new Earl styling we know today. Above: Had the original 'Vette been facelifted as first planned, it might have given way for '57 to this jazzy number clearly patterned on the 1955 LaSalle II roadster, another Motorama special. One exception: big, pointy front-bumper bullets, which seem to have been borrowed from Buick's '54 Wildcat II showmobile. The sweeping elliptical bodyside "coves" did appear on the production '56 Corvette, the one element retained from this interim proposal.*

Top: *Chevy began styling the third-series 1958 Corvette in 1955. Initial efforts took off from the Olds Golden Rocket being prepped for the '56 Motorama show. This studio shot shows how an early 'Vette-ized" GR looked next to a '56 Ford Thunderbird; an Austin-Healey and Mercedes-Benz 300SL Gullwing lurk behind.* Center: *Though the '58 'Vette design was frozen in August '56 (it's peeking in at the far right), Chevy evaluated this alternate face and a large, center rear-deck fin. Both elements came from the '56 SR-2 racers and a similar custom 'Vette built that year for GM president Harlow Curtice. Some believe these items may have been intended for factory racing programs.* Above left and right: *The 1963 Sting Ray coupe came together fast and early, as shown by this October '59 photo. Roof is more rounded here than on production '63s, reflecting work on the "Q-Corvette" of 1957, a more radically engineered design once proposed for 1960 but killed for cost reasons in favor of a more conventional car with styling patterned on Bill Mitchell's late-Fifties Stingray racing roadster.*

Top and center: *It's a '63 split-window Sting Ray coupe all right, but look closer and you'll see that doors, side windows, and wheelbase are all stretched to make room for "+2" seating. Photographed in late January 1962, this back-seat 'Vette was briefly considered for production with an eye to stealing some thunder from Ford's T-Bird, but the idea was* canned for being out of step with the Corvette's sports-car character. Above left and right: *GM knew this one as XP-895, but everyone else thought it the long-rumored mid-engine Corvette. Unveiled in 1972, it was built on one of two chassis from the earlier XP-882 midships project under Corvette chief engineer Zora Arkus-Duntov, which* broke cover with a 1970 one-off. Neither effort led to any future Corvette. However, a duplicate XP-895 body was later fabricated in Reynolds aluminum for publicity and research purposes, and the car survives in that form today as the property of Chevrolet.

Top left: *Though not called Corvette, the curvy one-off Astro II of 1968 fueled rumors that a mid-engine Chevy sports car was near. Developed as GM project XP-880, it was a follow-up to the previous year's Corvair-based Astro I, but carried a Corvette V-8 and conventional doors. The entire rear half of the body was hinged to tilt up for engine* access. Top right: *The Astro-Vette was Chevy's other big auto-show star in 1968, but it was merely an exaggerated version of that year's all-new "Shark" production design. Allegedly good aerodynamics were never proven.* Above: *It looks a bit like post-1982 Corvettes, but the one-off XP-898 of 1973 was actually built on the chassis of* Chevy's small four-cylinder Vega. Its *mission was to test feasibility of a new "sandwich" fiberglass body construction using a foam filler that could be varied in thickness to provide desired strength in specific areas. Though it looks a bit dated now, XP-898 would have been a great replacement for the '68-vintage "Shark" Corvette in, say, 1975.*

Aerovette and Friends: The Mid-Engine 'Vette

For one brief shining moment, General Motors honestly intended to build a mid-engine Corvette for public sale. The moment came in late 1977, just as "America's only true sports car" was about to celebrate its 25th anniversary. Regrettably, the decade-old "Shark" model had to carry the birthday banner, because the midships 'Vette wasn't slated until 1980. But enthusiasts wouldn't have minded the wait, for the car in question was a virtual clone of the stunning Aerovette, perhaps the most widely admired of the many mid-engine experiments with which GM had been teasing Corvette lovers since the late Fifties.

Those tantalizing exercises were owed to Zora Arkus-Duntov, fabled as chief Corvette engineer almost since the car's 1953 inception. After conjuring the open-wheel CERV I single-seater and envelope-bodied CERV II (the letters stood for Corvette Engineering Research Vehicle), Duntov turned to more passenger-oriented designs, beginning with the Astro II of 1968. Like the previous year's Corvair-based Astro I, this was a curvy, ground-sniffing two-seat coupe with a lift-up rear engine cover/cockpit canopy. It was also a remnant of project XP-880, a mid-engine effort that Duntov hoped would appear in showrooms for 1968. But GM decided to stick with traditional front-engine design and a little-changed 1963-67 Sting Ray chassis for that year's new "Shark" generation, thus rendering Astro II a dead end.

Undaunted, Duntov quickly turned to what would be the genesis of the Aerovette, project XP-882. Because his previous mid-engine proposals carried Chevy V-8s in longitudinal fashion, they required a costly, purpose-designed transaxle that not even vast Chevrolet could justify for a low-volume sports car. Here, Duntov tackled the problem by turning the engine 90 degrees and putting a stock GM Turbo Hydra-matic "end on" to it. The transmission was driven by chain from the crankshaft, and connected to a stock Corvette differential via a short driveshaft turning a right-angle at the front. Because the driveshaft had to pass through the sump, it was encased in a tube. If not an elegant solution, it was at least affordable.

Duntov's engineers built two XP-882s during 1969, an identical pair of swoopy fastbacks with an unfortunately blunt front but a dramatic louvered boattail, as on the experimental Mako Shark II of four years before. Yet almost on the day they were finished, John Z. DeLorean became Chevy general manager and canceled the program as impractical and costly. His decision stood only a year. When Ford announced plans to sell the Italian-built mid-engine DeTomaso Pantera through Lincoln-Mercury dealers, DeLorean ordered one XP-882 cleaned up for display at the 1970 New York Auto Show. But though car magazines were quick to proclaim that the mid-engine Corvette had finally arrived, GM never said anything about production.

Meantime, GM was working feverishly on its own rendition of the rotary-piston engine devised by Dr. Felix Wankel at Germany's NSU, having secured a manufacturing agreement at the behest of president Ed Cole, the legendary Chevy engineer who had lately become an ardent rotary advocate. Along with a Wankel-powered version of Chevy's small Vega, which would never materialize, Cole ordered up a sports car designed around the developing two-rotor GMRCE (General Motors Rotary Combustion Engine) then being eyed for production. Coded XP-897GT, this handsome little coupe had GM styling, but was built by the famed Pininfarina works in Italy. When displayed during 1973 with the prosaic title "Two-Rotor Car," "buff books" again hailed the advent of the mid-engine Corvette.

The previous year, DeLorean had authorized further work on the XP-882 chassis, as well as a new body from the corporate Design Staff under William L. Mitchell. Sufficiently changed to warrant a new project code, XP-895, this ended up looking a bit like the Two-Rotor from the sides, but carried a deeply inset "sugar scoop" rear window instead of flush glass. By early '72, a chance discussion with officials at Reynolds Metals Company prompted construction of a near-identical body in aluminum alloy, and in which form the XP-895 became the "Reynolds Aluminum Car." It, too, garnered lots of ink as the presumed next Corvette—and because its big-block 454 V-8 promised super performance against a svelte curb weight of around 3000 pounds.

As if all this weren't enough, the remaining XP-882 chassis was stripped of its V-8 and given a pair of Vega Wankels bolted together into a four-rotor, 420-bhp super-rotary. To make sure no one missed the change, Duntov persuaded Mitchell and staff to design yet another all-new body for what was called . . . the "Four-Rotor Car."

Exceptional aerodynamics was its overriding design goal. At just 44 inches high, the Four-Rotor tested out with a drag coefficient of only 0.325—better than many production cars of 20 years later. But its real triumphs were sumptuous sensuality and remarkable symmetry. Unlike most midships designs, the Four-Rotor gave no clue as to the location of its engine; indeed, its styling would have suited a front-engine layout just as well. And where most "aero" bodies had definite edges, this one cleverly disguised them. The result was organic, "all-of-a-piece," and nearly timeless—a triumph of surface over line.

Jerry Palmer, who would shape the sixth-generation 1984 Corvette, was among the designers who worked on the Four-Rotor car, which was first shown in late 1973. *Car and Driver* magazine thought it "the betting man's choice to replace the Stingray." But that winter brought the world's first energy crisis, which exposed the Wankel as a relative gas guzzler. With that, GM scrapped its rotary work and all plans for Wankel-powered cars.

Three years later, the Four-Rotor was still under a sheet in

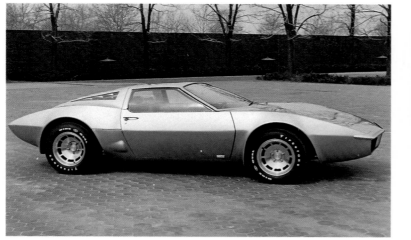

Top left and right: *Shot in the GM Design Staff viewing court just before its public debut in early 1970, the experimental XP-882 looked production-ready, thus fueling hopes that the next new Corvette would have a similar mid-engine design. It definitely looked like a Corvette, with overtones of the 1968-* vintage *"Shark" model in its low vee'd nose and four-lamp tail treatment. The car would have stayed under wraps, but was shown to counter Ford's announced sale of Italian-built DeTomaso Panteras. GM built two XP-882 chassis for evaluation, but only the first one had the bodywork shown here.* Above: *Built by* Pininfarina *to a GM design, the XP-897GT "Two-Rotor Car" appeared in 1973 as a showcase for GM's then-imminent Wankel-type rotary engine. Like the original XP-882, it was widely believed to be a precursor of the next-generation 'Vette.*

GM's Special Vehicles warehouse. Like DeLorean (who left GM in a huff during 1972), Mitchell had the car dragged out, this time to replace the double-Wankel with a Chevy 400 V-8. After changing the I.D. to "Aerovette," Mitchell lobbied for the car as the next Corvette. He usually got what he wanted, and GM chairman Thomas Murphy actually approved the Aerovette for 1980 production. Ironically, he might have been at least partly swayed by the imminent threat from the rear-engine DMC-12, the now-infamous sports-car effort of none other than John Z. DeLorean.

The process to productionize the Aerovette moved swiftly. A full-scale clay was ready by late '77, and tooling orders were about to be placed. The showroom model would have had a steel frame with Duntov's clever transverse driveline and probably a 350 V-8, which was then Corvette's mainstay engine. Transmissions would have likely been the usual four-speed manual and three-speed Turbo Hydra-Matic, and suspension would have come off the old "Shark" per Duntov's original cost-cutting aim. So despite its complex gullwing doors, the Aerovette wouldn't have cost a whole lot more to build than a front-engine 'Vette. Indeed, 1980 retail price was projected in the $15,000-$18,000 range. Best of all, the gorgeous styling would have survived completely intact. As Mitchell later confirmed: "The only difference between the Aerovette and its production derivation was an inch more headroom. Otherwise it was the same."

But once more, the mid-engine Corvette was not to be. There were several reasons. First, the project lost its two biggest boosters when Duntov retired in 1974 and Mitchell followed suit three years later. Ed Cole was gone by then, too. A further blow came from Duntov's successor, David R. McLellan, who preferred the front/mid-engine concept over a rear/mid layout for reasons of packaging, manufacturing economy, even on-road performance.

But the deciding factor was sales—or rather the likely lack of same. Though Porsche, Fiat, and other import makes had all proffered mid-engine sports cars for several years, none had sold very well in the U.S. Datsun, meantime, couldn't build enough of its admittedly cheaper front-engine 240Z—as GM bean-counters evidently observed. Simply put, the mid-engine was risky.

It bears mentioning that GM explored one other avenue at the same time as the production Aerovette. This was a midship *V-6* Corvette with running gear taken from the planned new 1980 X-body compacts. It was the same idea later applied to Pontiac's Fiero: a transverse front-drive powertrain plunked behind a two-seat cockpit to drive the rear wheels. The concept was hardly new, of course. Porsche, Lotus, and Fiat had all used high-volume, off-the-shelf components to create roadgoing middies—the "corporate kit car" formula that promised similar cost savings here. The contemplated engine was the now-familiar 60-degree 2.8-liter V-6 then in the works at Chevrolet. Styling was created by the Chevy Three Production Studio under Jerry Palmer, which sculpted clean, somewhat angular lines with Aerovette overtones.

But the mid/V-6 was doomed by the same factors that killed the Aerovette. It had other drawbacks, too. As *Car and Driver* later recounted: "A new front-engine/rear-drive Camaro had just been approved with [350-cid] V-8 capacity; there was no way a V-6 Corvette could continue as the flagship of the Chevy fleet without turbocharging and intercooling, and it would be tough to sell such a costly, high-tech alternative to management. At the same time, the corporation had yet to develop a transaxle that could withstand the torque such an engine would produce. In addition, GM had big plans for widespread use of its X-car components in future high-volume cars . . . limiting the availability of parts [for Corvette]." Exit mid/V-6.

With that, work toward a new front-engine design got underway in earnest during 1978. The result appeared five years later, in time for Corvette's 30th birthday. Mitchell later compared its styling to a "grouper." Tellingly, this sixth-generation 'Vette has yet to equal the sales performance of the old "Shark," encouraging hopeful types to anticipate yet another fling with mid-engine design. But those hopes were dashed by GM's 1993 announcement that the next-generation Corvette—to debut in 1996 as a '97 model—will continue the familiar front-engine rear-drive configuration.

Arguably more stunning than the Two-Rotor XP-897GT was the so-called "Four-Rotor Car" that appeared a bit later in 1973. Built on the first XP-882 chassis under the aegis of company design chief Bill Mitchell, it carried a pair of GM's experimental two-rotor engines bolted together into a 420-horsepower "super Wankel." A Corvette-like face and obvious high performance potential were taken as strong suggestions that GM was brewing a radical new Corvette for the late Seventies or early Eighties.

The Four-Rotor Car got a transplanted V-8 in 1976 to become the Aerovette, which came close to production four years later. GM design chief Bill Mitchell kept its original lines intact, however—not that there was reason to fiddle. A good-looker even now, it's a dynamic design

even when viewed from overhead (top). In profile (above), it displays a strongly triangulated "mound" shape, deftly balanced proportions, and artful surface detailing. "Gullwing" doors (center left) harked back to the original Mercedes

300SL coupe, but were articulated for easier operation in tight parking spots. Interior (center right) was more fully engineered than the show-car norm, another indication that the Aerovette was indeed a serious production prospect.

Missing MoPars: Never-Were Postwar Chryslers

Custom Club Coupe

Chrysler's wood-body Town & Country wagon was a modest 1941-42 success that led to postwar plans for a full non-wagon T&C line. To be included were six- and eight-cylinder two-door brougham sedans, convertibles, and four-door sedans, plus straight-eight hardtop coupe. The last (above left), created by grafting a metal roof onto the ragtop, would have preceded GM's first pillarless coupes, but Chrysler built only seven T&C hardtops, all in 1947. The broughams and six-pot convertible didn't make it either, but the straight-eight sedan (top) saw 2718 copies for '46, its six-cylinder sister some 4050 for 1946-48. For the all-postwar '49 line, T&C was reduced to just the eight-cylinder convertible, but Chrysler built one prototype '49 hardtop (above right) as a prelude to its production Newport models for 1950, when the T&C switched to that body style.

Though Dodge debuted a natty low-cost roadster for 1949, Plymouth could have had one, too. These mid-1946 photos show a near-final version of what was first proposed in 1944 as the "Interceptor." Unlike Dodge's initial Wayfarer model, Plymouth's "roadster" would have had roll-down windows (top) and a predictive lift-off top made of that new wartime wonder material, fiberglass (center right and above). Styling was updated from wartime designs (represented by scale models in some of the backgrounds here) for a more modern squarish look, but Plymouths wouldn't get flush rear fenders like this until 1953. Simple "face" with peaked horizontal bars (above left) contrasts strongly with the busy chrome latticework used on production '49s. Note the wide spacing between bumpers and body here, and the prominent vertical bumper guards.

Chrysler's Stillborn Sports Cars: Excitement From Exner

History must forever record the unfortunate 1989-90 TC by Maserati as Chrysler's first production "sports car." Which is a real shame considering the hot Dodge Viper that followed it—or, for that matter, the many sports and GT ideas Chrysler cooked up in the Fifties.

Virgil Exner headed Chrysler styling in those days. He's perhaps best remembered as instigator of the 1955 and '57 "Forward Look" cars that turned the company's red ink to black by bringing real excitement to its products for the first time in decades. But his influence was apparent much earlier. Teaming up with Italy's renowned Carrozzeria Ghia, Exner created a series of glamorous "idea cars," as Chrysler called them, starting with the four-door Plymouth XX-500 of 1950. Though not often appreciated, some of his sportier concepts came close to reaching dealerships.

Chrysler K-310 and C-200, 1951-52

Like most of Exner's Ghia-built specials, the dashing five-passenger K-310 coupe was designed in Detroit under Ex's eye. In this case, Ghia received a ⅜-scale clay plaster to guide construction of a full-size running prototype. The "K" stood for then-company president K. T. Keller, the "310" for the alleged horsepower of the 331-cubic-inch hemi-head V-8 beneath the hood, though the then-new stock version produced only 180 bhp.

No matter. The K-310 was stunning. Bulging integral rear fenders avoided period slab-sidedness, while prominently crowned front fenders emphasized a classic front with prow-style hood and headlamps recessed in scalloped nacelles astride a low, rounded, roughly triangular eggcrate grille. Full cutouts emphasized the wheels, which Ex seldom covered on any of his designs. Subtle two-toning delineated upper from lower body. The roof and deck were proportioned to accent the hood, which wasn't easy given the contemporary Chrysler Saratoga chassis with 125.5-inch wheelbase.

The K-310's warm reception prompted construction of a soft-top companion called C-200, unveiled in 1952. Both were strongly considered for showroom sale. As Exner later recalled, K.T. Keller liked the K-310: "He thought it was something they should promote . . . . Of course, it was also something into which they could put their Hemi engine. It was a perfect combination."

But the K-310 would be a never-was for the most basic of reasons: lack of money. Chrysler sales began to free-fall after 1949, and within three years the firm was outproduced by Ford for the first time since the Depesssion. But though plans for a limited run of "street" K-310s were shelved, Exner continued campaigning for a Chrysler-based sportster the public could buy.

Chrysler Special and D'Elegance 1952-54

Showmobiles have always been great attention-getters, with sales- and profit-boosting potential. This likely explains why Ex got the nod to do a series of suave GT coupes in the midst of Chrysler's cash-flow crisis of the early Fifties.

First up was the Special, which was built in two versions. The original premiered at the 1952 Paris Salon as a three-place fastback built on a cut-down New Yorker chassis (119-inch wheelbase). As a follow-up to the K-310/C-200, it sported similar "elements of Continental styling"—long-hood/short-deck profile, big wire wheels within full cutouts—but differed most everywhere else. Fenderlines were squared-up knife-edge types holding slim vertical bumperettes; headlights lived in prominent thrusting pods; the grille was an inverted trapezoid with horizontal bars. Also, bodysides curved less, and combined with a low roofline for a husky "masculine" air. Though handsome, the first Chrysler Special would remain one-of-a-kind.

So, too, the second version built in 1953 for C. B. Thomas, the head of Chrysler's Export Division, thus prompting the nickname "Thomas Special." Though similar to the '52 car, this mounted a stock 125.5-inch New Yorker chassis and measured 10 inches longer overall (214 total). Exner used the extra length to provide what we'd now term a notchback profile, with a normal trunk and external lid, plus four/five-passenger seating. There were also various detail changes, such as outside door handles instead of solenoid-activated pushbuttons.

In one sense, the Chrysler Specials were not dead-ends, for positive public reception prompted some 400 copies of a third version in 1954. This was dubbed GS-1, likely for "Ghia Special," though the styling was again Exner's. Overall appearance was somewhere between the two Specials. The main differences involved a larger and squarer grille, reshaped roof and fenderlines, and stock '54 New Yorker bumpers. All GS-1s carried the by-then familiar 331 Hemi V-8, linked to Chrysler's new fully automatic two-speed PowerFlite transmission. Sales were handled by Societe France Motors, Chrysler's French distributor.

Meantime, Exner returned to close-coupled fastbacks with 1953's Chrysler D'Elegance. This was more two-seater than three-place car, for its New Yorker wheelbase was trimmed to a tight (for the time) 115 inches. Though clearly evolved from the first Special, the D'Elegance was busier: gunsight taillights astride a dummy decklid spare, and a face much like the K-310's, right down to a prominently peaked bumper. Rear fenders were noticeably bulged, with leading edges dropped down from beltline level to near the rockers, where they continued to the front wheel arches as a horizontal creaseline.

If the Ghia-built D'Elegance looks familiar, it should. Though little appreciated for some years, Volkswagen acquired manufacturing rights to this design, which was then downscaled by Ghia to fit the chassis of VW's small, Thirties-era Beetle sedan. Germany's Karosserie Karmann was contracted as body supplier for what was introduced in late 1954 as the VW Karmann-Ghia—

Top and middle left: *The 1951 K-310 was the second of the Ghia-built Chrysler "idea cars" designed under the watchful eye of company styling director Virgil Exner. "Elements of Continental styling" were featured, according to* Chrysler, *but also several "classic" touches—like the dummy "toilet seat" spare tire outline—that would typify future Exner designs. Middle right and above: The C-200 appeared in 1952 as an open companion to the K-310. Also* built on a stock Chrysler Saratoga chassis, it shared the distinctive "gunsight" taillamps that would transfer virtually without change to the 1955-56 Imperial.

a gross misnomer, as the Italian firm had nothing to do with the original styling. It was one of the few times Ex didn't get the credit he deserved.

Dodge Firearrow I and II, 1953-54; Dodge Firearrow Sport Coupe, 1954; Dodge Firearrow Convertible, 1954

The Ghia-built Dodge Firearrow roadster first appeared as a two-seat mockup that rode atop a '54 Dodge chassis. Bright red and circumscribed by a dramatic gray molding that culminated up front in a handsome, blade-like bumper split by a single (and rather phallic) pod, Firearrow I carried dual headlamps and full wheel covers. Exposed exhaust pipes, two per side, rode low on the car's flanks. Inside, the seats were well-padded yellow leather adorned with narrow maroon piping. A wood-rimmed steering wheel brought an additional touch of Italian style.

Firearrow II, a modified version of the original car, retained the mockup's two-place seating and striking frameless windshield when it appeared in 1954. Riding on a 119-inch wheelbase and painted a subdued yellow, Firearrow II sported chrome-plated wire wheels instead of the previous full-wheel discs. The body molding was no longer a wraparound affair, but ended at the headlamps and taillights. In front, the dual headlights had become single units, and the original's gracious-looking split-bumper design was replaced by a more aggressive "mouth" horizontally bisected by an uninterrupted bumper. Five vertical design elements on the bumper gave the grille a toothy look.

Later in 1954, the two-seat Firearrow Sport Coupe appeared. As with the earlier roadster, the metallic blue coupe was essentially a '54 Dodge. Dual headlights returned, and now flanked a concave grille cut with narrow verticals. Crash protection front and rear was provided by modest bumperettes. A wraparound backlight gave the Sport Coupe a particularly rakish aspect. And the car went as good as it looked (with a modified engine, that is): Driving at Chrysler's Chelsea, Michigan, proving grounds in 1954, racer/flier Betty Skelton set a women's closed-course record at an impressive 143.4 mph.

The last of Virgil Exner's Firearrow series, the Firearrow convertible, arrived late in 1954. Despite being the series' first four-seater, it shared many styling cues with the Sport Coupe. The concave grille returned, though it now carried a grid treatment instead of the coupe's slim verticals. As for the convertible's leather interior—well, as it was a diamond pattern done in hard-to-ignore black and white, it was definitely an acquired taste. Additional sizzle was provided by the car's bright red body.

Happily, Exner's Firearrow series tickled the fancy of wealthy car enthusiast Eugene Casaroll, who purchased production rights to the design and teamed with engineer Paul Farago to create a practical road car. The result was the 1956-58 Dual-Ghia—proof that concept cars given cavalier treatment by the companies that commission them can sometimes be taken very seriously indeed by others.

DeSoto Adventurer I, 1954

The arrival of Chevy's two-seat Corvette in 1953 prompted this dashing one-off in 1954. Though visually related to earlier Exner specials, it mounted a '53 DeSoto chassis cut to a suitably

Top left: *Exner's 1953 Chrysler D'Elegance was a "2+1" with a single sideways rear seat and a cut-down New Yorker chassis. Essential shape was later scaled down and adapted for the VW* Karmann-Ghia. The tastefully plush interior (top right) of the '53 "Thomas Special" (above left) was just one highlight of this one-of-a-kind Exner exercise, a modified version of the 1952 *Chrsyler Special. Above right: The Dodge Firearrow II was a modified 1954 version of the original Firearrow roadster. Note the frameless windshield. The car is now owned by Chicago-area collector Joe Bortz.*

Top and middle left: *Exner lobbied hard for a production version of the racy 1954 DeSoto Adventurer, and though it came closer to approval than any of his other specials, Chrysler management just didn't have the courage. Adventurer could seat four, despite its closely coupled styling. Aggressive side exhausts* foreshadowed a feature of the far-distant Dodge Viper. Small rear hatch allowed access to the spare tire, but luggage space was evidently next to nil. Middle right: *The 1955 DeSoto Adventurer II was more Ghia than Exner. Though larger than the original Adventurer, it had seats for only two. The front bumper was* dispensed with; up back, the plastic backlight was retractable into the rear deck. Above: *The '54 Firearrow Sport Coupe was another variation on the Firearrow theme. Note the token bumperettes and wraparound backlight. The car is another gem in the collection of Joe Bortz.*

sporty 111-inch wheelbase. Despite the close-coupled coupe styling with no rear side windows, the Adventurer could hold four in comfort. Highlights included a new iteration of the inverted-trapezoid grille, functional side exhausts, another quick-fill fuel cap, the usual chrome wires wearing "wide whites," off-white paint, and minimal bright accents. The interior was swathed in black leather with white piping, and satin-finish aluminum set off a dashboard with a complete bank of circular gauges.

Exner tried very hard to get the Adventurer approved for limited production. But as Maury Baldwin, one of his staffers, later recalled, "Management at that point was very stodgy. A lot of people attributed it to the old Airflow disaster. They were afraid to make any new inroads." Ex himself later said that the Adventurer came closer to production than any other Chrysler special to that time: "If it had been built, it would have been the first four-passenger sports car made in this country . . . . Of course, it had the DeSoto Hemi [a '53 stock 273 with 170 bhp]. It was my favorite car always . . . " A second DeSoto-based exercise, the '55 Adventurer II, was mainly Ghia's work and never a serious production prospect.

Chrysler Falcon, 1955

The two-seat Falcon was the closest Chrysler came to a classic sports car before the Dodge Viper almost 40 years later. Many still think it was the one Exner special that should have been built for sale. After all, by the time it appeared, Ford had introduced the '55 Thunderbird and Highland Park sales were fast recovering, so a Mopar reply to both the Corvette and the "personal" Ford would have been quite timely.

Chrysler must have thought so too, for it built three Falcons. Though they differed somewhat in details, all rode a 105-inch wheelbase, comparable to the 102-inch T-Bird and Corvette. The ruggedly handsome styling was mainly the work of Maury Baldwin and still looks good today, especially the big heart-shaped eggcrate grille and rakish side exhausts. Even the trendy Fifties fins and wrapped windshield don't seem particularly dated.

The Falcon carried a 170-bhp DeSoto Hemi, like Adventurer I, but bypassed its old "fluid-torque" semi-automatic transmission for fully automatic PowerFlite, which was controlled—none too positively—by a wispy floor-mounted wand. Unitized steel construction hinted at things to come from Chrysler, though it pushed curb weight to a portly 3300 pounds.

Still, Falcon performance was perfectly adequate. In a brief road test of the only known survivor some years back, a contributor to this book clocked 0-60 mph in 10 seconds flat, about 115 mph all out, and a standing quarter-mile of 17.5 seconds at 82.0 mph—all more than adequate for 1955. Mileage? About 15 mpg.

Despite its heft, the Falcon had beautifully balanced handling and easy yet precise steering of the sort virtually unknown in period Detroiters, especially Chryslers. Its one real drawback was lack of top-up headroom due to its very low windshield, though that would have been fixed for production.

Which, of course, didn't happen. Though the Falcon would

have been a strong competitor for Corvette and Thunderbird, with arguably superior refinement and performance, it was doomed by the minuscule sports-car market of the time. Also, Chrysler likely felt it really didn't need such a car so long as overall sales were good—which they weren't after 1957.

Plymouth XNR, 1960

While Ford's Falcon began running away with the compact market, Exner was transforming a Valiant into his most radical idea car of all, the Plymouth XNR (the meaning of the initials is obvious). Representing the peak of his enthusiasm as head of Chrysler design, its imaginative "Asymmetrical Styling" was bold, to say the least—especially for a Plymouth.

The idea was to emphasize the left-side driver's position. This was achieved by a prominent offset hood scoop faired back into the cowl, which was topped by a squat, racer-style windscreen; the line then carried back through a large headrest/tailfin. Quad headlamps nestled in a big mesh-filled bumper/grille roughly oval in shape. The passenger seat was normally covered by a metal tonneau, but a small fold-flat auxiliary windshield was provided should a co-pilot be aboard.

Though XNR rode Valiant's tidy 106-inch wheelbase, prominent overhangs stretched overall length to 195 inches. Height was just 43 inches to the top of the fin. Power came from the hairiest version ever developed of Valiant's 225 Slant Six, which pumped out 250 horsepower—1.11 bhp per cubic inch. "We took [XNR] to the Proving Grounds and had a professional drive it," Exner said later. "He lapped at 151 or 152, which wasn't bad for that time." Or ours, come to that.

What excited sports-car fans—and prompted rumors of imminent sale—was the XNR's engine: Chrysler's first high-performance six and only the second such U.S. powerplant after the fabled Hudson Hornet six. As a production sports car the XNR would have been unique; in racing guise it would likely have trimmed most anything in its displacement class. But again, Chrysler decided there was just no market; even if there had been, the styling would likely have seemed just too far out to sell well. Finally, Exner's abrupt firing in 1962 killed any chance the design might have had for being refined into something more practical for production.

Incidentally, the XNR was one Exner special *not* built by Ghia, though the Italian coachbuilder did manage something similar on its own a bit later. Alas, its Valiant Assimetrica had none of the XNR's flair, and never went beyond the one-off stage.

It's a pity the sleek K-310, the lovely Adventurer, and the burly Falcon didn't make it onto America's roads, for it's easy to imagine how pleasing they would have been. But at least Exner (and the rest of us) had the satisfaction of witnessing the Firearrow-inspired Dual-Ghia.

Whatever their ultimate fate, Chrysler's stillborn sports cars remain among the best examples of Virgil Exner's inimitable legacy. As a designer he was as unique as any of his creations and the time in which he flourished—a younger, more innocent age we'll never see again.

Opposite page, top and center left: *Executed under Exner's aegis by staff designer Maury Baldwin, the 1955 Chrysler Falcon was as close as the corporation got to a production sports car before the 1991 Dodge Viper. Three were built by Chrysler's Advanced Styling Studio, each with a 105-inch wheelbase and 170-bhp DeSoto Hemi V-8. Styling details differed among them, and only one is known to have suvived. The Falcon's* most noteworthy feature lay beneath the car's skin: unit construction with an integral cellular platform frame. Although the Falcon had the look of a posh boulevardier, it was envisioned as a potent performer, as well. The convertible top was operated manually, and could be stowed beneath a folding lid located behind the seat. Handsome, laudably subdued styling looks good nearly 40 years later. Road manners were *reportedly impeccable, performance at least adequate. Exner later "gave" the Falcon name to Ford for its new 1960 compact.* Opposite, center right and below: *Exner was enamored of asymmetric design by 1960, when his Valiant-based XNR appeared. The car was likely intended to be a preview of the similar but far more subtle styling then being planned for Chrysler's 1962 showroom models. The car could seat*

two, but was best suited for a single occupant, the driver. Thus the huge portside headrest-cum-tailfin, which was intended to emphasize the driver while harkening back to late-Fifties racing-car design, exemplified by the likes of Jaguar's D-Type and XKSS. The driver sat behind a dramatically curved "personal" windshield; a smaller, fold-down windscreen was available for the protection of a passenger. Additionally,

the passenger sat somewhat lower than the driver—a design touch intended to minimize the negative effects of the wind. The frame of the XNR's grille was constructed of heavy-duty materials, and doubled as the car's front bumper. The "X-motif" rear bumper was a visual reminder of the car's name and essentially asymmetric nature. The interior was finished in black leather and aluminum. Of the car, Exner remarked he

was "striving to avoid the static and bulky, which is ugly and not what an automobile should look like. The goal is to try for the graceful look, with a built-in feeling of motion. The wedge shape expresses the function of automobiles because it imparts a sense of direction." Unlike earlier Exner specials, the XNR (the name, of course, allowed Exner a self-referential giggle) was built by Chrysler, not Ghia.

77

Chrysler's Turbine Cars: Promise Unkept

Many still remember the Chrysler Turbine Car, the bronze-colored hardtop that looked like a Thunderbird and whined like a banshee. Unveiled in 1963, it seemed to promise that turbine power would soon be in your driveway—as indeed it *was* for the couple hundred people who drove the 50 Turbine Cars built in a three-year consumer test program. But though Chrysler came closer than anyone to perfecting a practical turbine automobile, events rendered the futuristic powerplant just so much hot air, and its promise went unkept.

The gas turbine relates to the jet engine, patented by England's Frank White in 1930. Rolls-Royce speeded the development of jet fighter planes, which came too late to significantly affect World War II but proved decisive in Korea. Between those wars, America's Big Three automakers began working on aircraft turbines (Chrysler built a turbo-prop for the U.S. Navy Bureau of Aeronautics in 1948), but didn't begin thinking about roadgoing applications until the mid-Fifties, perhaps inspired by Rover of Britain's 1952 experimental turbine car. Of the Detroit producers, only Chrysler seemed as serious about turbine-powered automobiles; Ford and GM put most of their turbine emphasis on trucks.

Like a jet, the turbine's basic element is a wheel ringed with blades or vanes; a fuel/air mixture flows past the vanes, causing the wheel to rotate and produce power. In many designs, including most of Chrysler's, this "power turbine" also ran a "first-stage turbine" linked to a compressor; the latter, of course, squeezed the mixture for firing, which was accomplished by a spark plug-like device called an igniter.

This simplicty appealed greatly to Chrysler and others because it meant fewer parts, which implied less maintenance. For owners, the turbine promised operating smoothness unknown in reciprocating engines, because it produced rotary motion to begin with. Other attractions included near instant warm-up (and available engine heat in winter), dependable cold-weather starting, the ability to run on a wide variety of fuels (Chrysler claimed the turbine could gulp everything from peanut oil to Chanel No. 5), negligible oil consumption, and no need for antifreeze.

Offsetting these pluses were four big minuses: high internal heat (upwards of 2000 degrees Fahrenheit); operating traits better suited to steady speeds, as in aircraft but typically not cars; no inherent "engine braking," vital on the road; and high oxides of nitrogen (NOx) emissions. Nevertheless, Chrysler would solve or at least minimize all of these problems over some 25 years of development.

High heat was the biggest problem. To attack it, Chrysler engineers under George Huebner, Jr., soon known as Highland Park's "Mr. Turbine," developed what they termed a "regenerator." This was essentially a rotating heat exchanger that removed exhaust-gas heat to reduce internal temperature and boost fuel

mileage above what it would be otherwise. Also studied early on were improved operating flexibility and development of materials resistant to ultra-high temperatures.

By 1954, Chrysler was ready with its first automotive gas turbine. Dubbed "CR1" and rated at a modest 100 horsepower, it was installed in a stock-looking '54 Plymouth Belvedere hardtop that ran successfully at the opening of Chrysler's new Chelsea, Michigan, proving grounds. A similar engine went into a '55 Belvedere four-door.

Improvements followed thick and fast. A modified CR1 with more exotic metallurgy powered a '56 Belvedere "Turbine Special" sedan on a cross-country run that year. It performed well except for returning only 13 miles per gallon, excessive even for those times. Next came a more efficient CR2 with about 200 bhp, aided by cheaper but sturdier alloys that better withstood heat and oxidation. This powered a second '56 Plymouth sedan as well as a 1957-58 version before making its public debut in a '59 Fury-based Turbine Special hardtop sedan. That car turned in a more creditable 18 mpg on a 1200-mile demonstration run from Detroit to Princeton, New Jersey.

The third-generation CR2A was ready by 1960. Its two big advances were a pivoting fuel nozzle mechanism and first-stage turbine vanes that could assume one of three angles depending on throttle position. Together, these provided a measure of engine braking, plus stronger acceleration; they also greatly reduced the irritating "throttle lag" that had plagued earlier versions. Where the CR1 needed a full seven seconds to go from idle to full power output, the CR2A took only 1.5-2.0 seconds.

CR2As were initially installed in three different 1960 vehicles: a near-stock Plymouth Fury hardtop, a 2.5-ton Dodge truck, and the amazingly befinned TurboFlite, which wasn't shown until 1961. The last was designed by Maury Baldwin, who later termed it the final "Virgil Exner" show car: "We incorporated a lot of interesting things in it. Entrance-wise, the whole cockpit above the beltline lifted to admit passengers. Mounted between the fins was a deceleration flap, such as used on racing cars. The headlights were retractable. The car was built by Ghia; we did a ⅜-scale model and then full-size drawings. It was probably one of the best engineered show cars we ever did."

After further development, the CR2A was fitted to a quartet of 1962-model hardtop coupes: two Dodge Darts and a pair of Plymouth Furys. One of the Turbo Darts, as they were called, traveled from New York to Los Angeles on another durability run, scoring better fuel economy than the piston-powered "control" car traveling with it and taking less time than the '56 Turbine Special. The '62 turbines were also shown at various Dodge and Chrysler-Plymouth dealers, and even went to Europe for track demonstrations at Montlhery in France and Silverstone in

Top left: *A '54 Plymouth Belvedere hardtop was the testbed for Chrysler's first car turbine, the CR1. Project engineer George Huebner, Jr. (on right) was Highland Park's "Mr. Turbine."* Top right: *Another period publicity photo emphasizes the '54 Turbine Special's cool-running nature despite the high internal heat inherent in turbines.* Middle row, left and center: *A '55*

Plymouth sedan was used to test refinements to the CR1 engine. The man checking under the hood is George Stecher, chief test driver for most of Chrysler's 25-year turbine development effort. Middle row right: *Stecher applies finishing touches to the final CR1 Turbine Special, a '56 Plymouth that managed only 13 mpg on a coast-to-coast run that year.* Above left: *Improved*

CR2 turbine was publicly unveiled in this '59 Plymouth Fury, which averaged 18 mpg on a 1200-mile demo run. That's Huebner again, on the left. Above right: *An interim test-turbine Plymouth was this stock-looking '58, which also carried the second-generation CR2 engine but wasn't publicly displayed. Note the four central exhaust outlets below the bumper.*

England. Buoyed by the uniformly favorable reaction to all this, Chrysler decided to build 50 turbine-powered passenger cars for consumer evaluation.

The result was that now-famous bronze hardtop, unveiled in May 1963. It was designed expressly for the consumer program by Elwood Engel, who had replaced Virgil Exner as head of Chrysler Styling two years before. Engel had come over from Ford after working on the 1961-63 Thunderbird, and his work on that car was clearly evident here. In fact, the resemblance was so strong that some referred to the Turbine Car as the "Engelbird."

Of course, there were many differences even apart from the powerplant. At 110 inches, Turbine Car wheelbase was three inches trimmer than the T-Bird's, and styling was unique at each end. The front was simple, if unfortunately blunt, but the back was wild, with deep "boomerang" cavities holding large, angled taillights astride backup lamps in big "turbine-styled" housings. Headlamp bezels and wheel-cover centers had a similar rotary-blade motif. All the "consumer cars" wore a black vinyl roof covering to contrast with the "Turbine bronze" paint. Interiors were also done in bronze, and offered seating for four on individual vinyl-covered seats, plus a full-length cylindrically shaped center console. Other amenities included power steering, brakes, and windows, plus modified TorqueFlite automatic transmission, radio, air conditioning, and appropriate instruments like tachometer and turbine-inlet temperature gauge.

Under the hood sat Chrysler's latest fourth-generation gas turbine, designated A-831. Its chief innovations were constantly variable first-stage turbine vanes, again controlled by throttle position, and twin regenerators rotating in vertical planes, one on each side of a central burner. Smaller and lighter than the CR2A, the A-831 was also quieter and more responsive, with throttle lag reduced to 1.0-1.5 seconds. Maximum-output engine speed after gear reduction was 4680 rpm, versus the CR2A's 5360. Horsepower was down 10, to 130, but torque went from 375 to a mighty 425 pounds-feet. In other respects, the Turbine Car was utterly conventional, though its TorqueFlite automatic had no torque converter (not needed), and there was "Unibody" construction per Chrysler practice.

Because of the small number involved, the 50 "production" Turbine Cars plus the five prototypes (three of which differed in roof/paint schemes) were contracted to Ghia in Italy, which could build them for less money than Chrysler. Meanwhile, Highland Park was swamped with over 30,000 inquiries after announcing the program, which called for each "consumer representative" to use a Turbine Car for about three months. Ultimately, the field test included residents of all 48 continental states ranging in age from 21 to 70. A total of 203 people drove Turbine Cars between 1963 and early '66.

The results were in by the following year, when Chrysler issued a report stating that little or no maintenance had been required compared with comparable piston-engine cars. However, the firm never divulged mileage figures, which were apparently embarrassing. One out of four drivers complained about gas guzzling (they must have been out of peanut oil), and one out of three groused about throttle lag. But there was also plenty of praise, especially about the vibrationless operation and snazzy styling.

Though Chrysler never released one to a journalist, writer

Above left: *Huebner inspects the further-improved CR2A turbine in a modified 1960 Plymouth Fury, one of three turbine showcase vehicles Chrysler built that year. This version advanced the state of turbine technology with a new* pivoting fuel nozzle and three-position first-stage turbine vanes whose angle varied by throttle position. Horsepower was about 140. Above right: *A pair of Dodge Turbo Darts (upper) and two Plymouth Turbo Furys were built to tout* Chrysler's turbine program in 1962. One of the Darts ran coast-to-coast in less time than the '56 Turbine Special and scored better mileage than a conventional piston-powered traveling companion. All four toured dealers, and a couple were demonstrated to journalists in Europe.

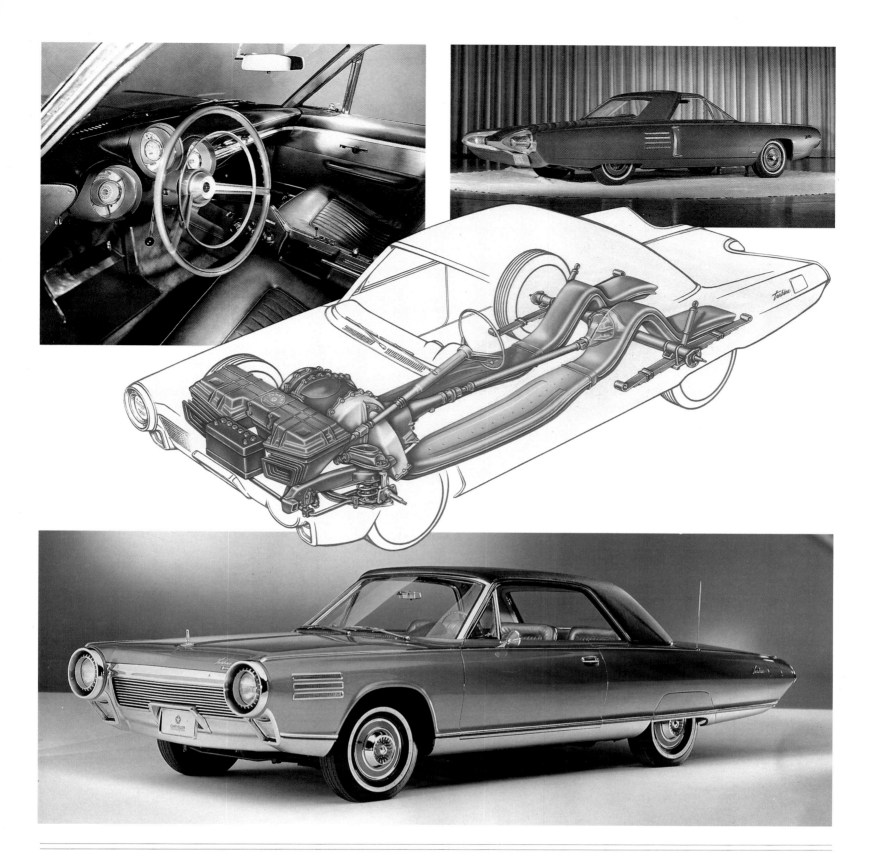

Top left: *Chrysler's 1963 "consumer" Turbine Car was rather understated inside, but cylindrical center console was striking. Brushed-metal door trim accented bronze-colored upholstery and carpeting.* Top right: *Though it ended up looking quite orthodox, the '63 Turbine Car might have been far more radical, as this mockup suggests. Two-passenger layout with abbreviated* landau-style roof was likely vetoed for more practical appeal among the car's consumer testers. Abandoning the bodyside scoop treatment was a blow for good taste. Center: *A typical bit of factory PR artwork reveals the Turbine Car's mechanical package, suspension— and huge exhaust ducts, necessary to keep exiting heat at reasonable and safe levels.* Above: *The '63 Turbine Car was* once called "Engelbird" for its general similarity to the 1961-63 T-Bird designed by Chrysler styling chief Elwood Engel just before he came over from Ford. All 50 of the "consumer" cars wore Turbine Bronze paint and black vinyl roof coverings. Note the engine-related "blade" motif on the wheel covers and around the headlamps.

John Lawlor managed to drive a Turbine Car toward the end of the consumer test. He, too, was impressed by its smoothness, but annoyed at the relative lack of engine braking—and at the throttle lag, which he reported as the claimed 1.0-1.5 seconds. His mileage also disappointed—just 11.5 mpg—though he usually ran on cheap kerosene. Lawlor did laud acceleration, which at under 10 seconds 0-60 mph was sparkling, especially for a 4100-pound car. So much for rumors that turbines couldn't be quick.

Yet even as the Turbine Cars were being passed around, Chrysler was moving ahead, testing a fifth-generation engine (A-875) in a '64 Plymouth. But that would prove short-lived, for a sixth-generation design was ready by 1966. Fitted to a Dodge Charger fastback and Coronet 500 hardtop, this offered improved engine braking, plus a novel split accessory drive by which ancillaries like the power steering pump were driven directly from the power turbine; the compressor turbine continued running the fuel pump and other engine-related systems.

Chrysler had actually planned to build 500 turbine-powered Coronets for retail sale, but nixed the idea because of imminent new government emissions standards. Still, that one '66 hardtop

Top: The '63 Turbine Car was wildest at the rear with huge "boomerang" taillamps and heavily sculptured deck. Finned cylindrical backup-lamp housings were one of many "turbine" styling symbols. Above left: Chrysler was up to a sixth-generation automotive turbine by 1966, when this Dodge Coronet was converted as a testbed for it. Chrysler almost built 500 such cars for public sale, but had second thoughts when new federal emissions laws loomed. Above center: This stock-looking Coronet sedan was one of several conversions tested in 1973. Above right: The seventh and final Chrysler turbine-engine design powered a pair of Dodge Aspen sedans that were stationed in D.C. under contract terms with the Energy Research and Development Adminstration. Both cars ran flawlessly.

hung around Chrysler Engineering until 1972, when the turbine tale took a somewhat surprising turn.

Though NOx emissions remained a thorny problem, the newly created Environmental Protection Agency (EPA) was persuaded—partly by a sales pitch from George Huebner—to give Chrysler $6.4 million to continue turbine development. Besides NOx control, specific objectives of the grant were to increase mileage, lower manufacturing costs, and provide at least comparable performance and reliability relative to "conventional piston-powered *compact-size* American cars" [emphasis added].

After tests with a trio of 1973 mid-size Dodge/Plymouth sedans, Chrysler unveiled a seventh-generation turbine. Though it reverted to a single regenerator, it boasted more precise electronic fuel control. Initially installed in a pair of '76 Dodge Aspens, this engine also powered a one-off T-roof coupe, basically a '77 Chrysler LeBaron with knife-edge front fenders, hidden headlamps, and slim vertical grille. Horsepower was only 104 versus the sixth-generation's 150, but this newest turbine ran somewhat hotter, so 125 bhp was available via water injection at the compressor inlet and repositioned inlet guide vanes.

Next, Chrysler landed a similar contract (along with GM and Ford) from the Energy Research and Development Administration (ERDA), which was later combined with several other agencies into today's Department of Energy (DOE). Still seeking turbine perfection, engineers soon virtually banished throttle lag, brought hydrocarbon and carbon-monoxide emissions within prevailing statutory limits, and attained fuel economy that approached that of comparable piston engines. Per contract terms, Chrysler stationed its two turbine-powered Aspens in Washington, D.C., where they ran flawlessly.

But by then it was 1979, and lower NOx levels still seemed impossible. Worse, Chrysler was racing toward bankruptcy, and a deep new recession was triggering federal program cuts all over. With that, the DOE withdrew funding in early 1981, and Chrysler soon abandoned turbine research entirely after more than a quarter-century and over $100 million of its own money, plus $19 million from taxpayers. In an eerie echo of the way it all began, the very last turbine car built was a near-stock-looking 1980 Dodge Mirada.

It's unfortunate things ended when they did. According to one project official, left stillborn was an eighth-generation turbine designed, ironically enough, for Chrysler's all-important new front-drive K-car compacts and their future derivatives. With a single turbine shaft (versus two), electronic fuel delivery, and a projected 85 bhp, it would have been the simplest turbine yet, and likely the cheapest to build in quantity. There were also hopes that a new variable-geometry burner would be the long-sought answer to NOx. But time and money had run out, so this engine went no further than blueprints and a foam mockup.

Fortunately, Chrysler showed a sense of history about the Ghia-built Turbine Cars, coughing up enough cash to save 10 from the torch. The rest were cut up under the watchful eye of U.S. Customs. They had to be. Import duty on these "foreign" cars had been waived only for purposes of the testing program; once that

ended, Chrysler had the choice of either returning them to Italy or paying considerable sums to keep them on American soil. Of the 10 that were saved, nine have been accounted for. Chrysler still has three; the remaining six have gone to various museums.

Turbine power now has as much relevance to our automotive future as rumble seats and tailfins, especially with the advent of draconian legislation mandating "zero-emissions vehicles"—a.k.a. electric cars. At least we have Chrysler's turbine tale and a few of its artifacts to remember a future that almost was, but in the end never could be.

Top: *Chrysler's last truly different-looking turbine experimental was this modified '77 LeBaron coupe. Front end foreshadows the face of the 1981-83 Imperial.* Center: *Looking a bit forlorn on the grounds at Highland Park, this*

stock-looking 1980 Dodge Mirada was Chrysler's last turbine-powered car. It was turned over to the government along with other project materials when Chrysler abandoned turbine research in early '81, due to lack of money. Above:

Test driver George Stecher in 1963 with one of the Ghia-built Turbine Cars and the Chrysler turbine engines devised to that point. Chrysler paid duty to save 10 of the 50 bronze hardtops from the crusher; nine still survive at last report.

Continental Mark II: What Didn't Happen for '58

Conventional wisdom has long held that the Continental Mark II lasted only two years because Ford Motor Company lost $1000 on each one sold. Actually, the Mark II was never intended to make money—and that was the problem.

It was born in 1952 to answer requests for a modern successor to the classic first Continental line of 1940-48, but it was also part of a bullish expansion plan. With Ford just returned from the financial brink and the company's future seemingly boundless, management decided to take on giant General Motors by going from two divisions to five: Ford, a separated Lincoln-Mercury, Continental, and one to handle a second medium-price make that ultimately emerged as Edsel.

Creation of the Mark II was assigned to a new Special Products Division (which later became Continental Division) under William Clay Ford, younger brother of company president Henry Ford II. Styling, chosen in 1953 from 13 different proposals, was the work of a stellar in-house team comprising John Reinhart (fresh from Packard), Gordon Buehrig (of Auburn-Cord-Duesenberg fame), and young Robert Thomas, with an assist from Thirties coachbuilding great Raymond H. Dietrich. The result was an elegant close-coupled hardtop coupe, impeccably made and completely free of period fads. In fact, many still rank the Mark II design as one of Detroit's all-time best.

Appearing for 1956 on a 126-inch wheelbase, the Mark II was priced at $10,000—breathtaking for the day, but reasonable given an unusual amount of hand labor and high luxury content. Indeed, air conditioning was the sole option ($740). Everything else was standard, including full power equipment.

But grand though it was, the Mark II never really got its chance. By the time it arrived, the ebullient executives who conceived it had been replaced by sober accountant types for whom no car was sacred unless it made money. Thus, soon after the Mark II was unveiled to critical acclaim, Ford decided to proceed with more profitable Lincoln-based models for 1958. The Mark II thus disappeared after '57 with virtually no change save a horse-power increase to 300. Total production ended at 1769.

Left behind were plans for a three-model 1958 Mark II line anchored by the familiar hardtop coupe. Only one styling change was contemplated: a mild frontal update with stacked quad headlamps (then all the rage), no vertical grille bars, and a mildly reshaped bumper. Not that there was reason to tamper. As John Reinhart said later, the Mark II "was so perfect a design that we felt it could go as long as 10 years."

More intriguing was Buehrig's suggestion for a long-wheelbase four-door hardtop called Berline ("sedan" in French). Evolved through numerous renderings and clay models, this used basic Mark II styling but looked sufficiently more "important" to be chauffeur-driven should the occasion demand. Differences

began with a slightly taller grille flanked by stacked quad headlights, as planned for the coupe, but with the lower lights nestled in the front bumper. Bodysides bore no decoration save ribbed rocker-panel accents, a coupe-style "character line" running full-length just below the belt, and matching lower-body crease that started as a vertical line just behind the headlamps.

The third model contemplated was a two-door retractable-hardtop convertible that was *not* inspired by the '57 Ford Skyliner. The "retrac" was actually engineered at Special Products in 1954 as the sole Mark II model, only to be rejected when program costs threatened to spiral out of sight. The idea wasn't passed on to Ford Division until after the decision to go with cheaper Lincoln-based '58 Continentals.

Reinhart and company believed each of the planned '58 Mark IIs would become "a 'classic' in its own right. But we got stopped at the gate." What stopped them was a Mercury cost expert sent in to make the Mark profitable. As a result, the production '58s were downgraded far below the proposed trio of Mark II "line extensions." However, the "Lincolnize" decision wasn't made for six months, during which time management considered holding for a while with just the Mark II coupe to see how sales would fare. Sadly, the Berline and retrac were shelved along the way.

Meanwhile, Continental Division commissioned a custom Mark II convertible coupe as a gift for Mrs. W. C. Ford. Built in 1957 by the famed Derham Body Company of Rosemont, Pennsylvania, it had a soft top with very wide rear quarters as on the original Continental cabriolets. A second convertible was later cobbled out of a Mark II coupe by a private party in Florida. The Derham car, at least, still exists, and testifies that an open Mark II would have looked great. Regrettably, Ford never thought of selling copies.

What it did sell for '58 was the giant Mark III, a mildly restyled version of that year's blocky new unibody Lincoln on the same 131-inch wheelbase. Styling was nowhere near as graceful as the Mark II's—the "slant-eye" face was particularly jarring—but buyers liked it well enough given vastly reduced prices: around $6000 for a choice of convertible, two- and four-door hardtop, or four-door sedan.

The Mark III turned a small profit on sales of 12,550, but Dearborn's expansive "divisionalization" dream was dead, a victim of the '58 recession and Edsel's abject failure. With that, Continental ceased to be a separate make, and Continental Division was folded into a short-lived Mercury-Edsel-Lincoln unit, which then became just Lincoln-Mercury (again).

After 1960, all Lincolns were Continentals until the new Mark III hardtop coupe of 1968. Never again would Ford consider anything quite like the Mark II Berline or retractable. A pity, for they would have been glorious.

Top row: *A retractable hardtop convertible was devised in 1954 as the sole Mark II body style and led to the 1957-59 Ford Skyliner. The idea was briefly considered as one of three models in a 1958 Mark II line, but was axed for cost reasons.* Center left: *The elegant Mark II would have worn this subtle four-headlamp facelift had Ford not dropped its super-luxury liner after 1956-57. The original vertical grille bars are missing on this hardtop mockup.* Center right: *To cut costs, Ford decided to drop the Mark II for a '58 Mark III cloned from that year's new Lincoln. This grille workout suggests designers first tried to retain as many Mark II cues as possible.* Other photos: *More views of the same early Mark III proposal. Ungainly taillamps and humped trunklid were wisely discarded, but lower rear fender trim would appear in modified form on standard '59 Lincolns. The 1958-60 Mark's reverse-slant rear window was still to come when this clay model was sculpted.*

The Last DeSoto and Other Never-Were '62s

DeSoto died quietly on November 18, 1960 after 32 mostly successful years. The reasons? Continuing Chrysler Corporation losses and high development costs for the new Valiant compact. The last DeSotos—a token pair of garish '61 hardtops—were little more than thinly disguised '61 Chrysler Windsors and numbered a paltry 3034—this from a make that had topped 125,000 units as recently as 1957.

But tucked away in the Highland Park styling studios was a brand-new 1962 DeSoto embodying a radical new design philosophy. Actually, all of Chrysler's full-size cars were to be completely redesigned for '62 in a top-to-bottom corporate overhaul known as the "S-series" program. As we know, only the Dodge and Plymouth members of this family made it to showrooms, and even they were drastically changed from original plans.

The S-series began taking shape in late 1958 under Chrysler design vice-president Virgil Exner. Its inspiration was his XNR show car, then in preparation, with similar long-hood/short-deck proportions and prominent blade-type fenderlines, elements slated for production premiere on the 1960 Valiant. After initial work with ⅜-scale models, a full-size "theme" clay was completed by May 1959 and wheeled under the styling dome. A convertible mocked up as a DeSoto, it represented a striking new direction for Exner: smaller, dramatically altered in profile, completely finless, and with a truncated rear deck in complete contrast to contemporary long-tailed Ford and GM cars.

Fins had made Chrysler the industry's styling pacesetter in 1955-57, but Exner knew they were already passé. With the S-series, he reached for the next "Forward Look"—literally—by shifting the focus from rear to front. Other features of that May mockup included vee'd bumpers, a slight beltline kickup just behind the doors, and a center vestigial fin (which survived only on the '62 Plymouth). Also featured were curved side glass (pioneered by Chrysler on the '57 Imperial), a steeply angled windshield, and long bladed fenders *a la* the XNR that stopped abruptly just before the B-pillar. Aft quarters wore a high-set concave rhomboid that tapered smoothly back to match rear-deck contour.

Somehow, management eagerly approved this mockup over more conventional clays from production studios (one surprisingly Mercury-like), leaving the individual division designers to evolve their '62s from it. Since Chrysler and DeSoto would share bodyshells and many outer panels as usual, DeSoto studio chief Don Kopka worked closely with his Chrysler counterpart, Fred Reynolds.

For both models, the theme car's rear-quarter treatment was quickly modified to harmonize better with the tapered deck, producing what junior stylists dubbed "the chicken wing." Front bumpers acquired dropped center sections because Exner wanted something other than straight bars. DeSoto designers ultimately revived their make's triple-taillight motif of 1956-59, though in horizontal instead of vertical format. The borning Chrysler was given large, single wraparound units.

Both makes carried four headlights in individual chrome bezels, arranged in slanted vertical pairs, as on the production '61s. Grilles were topped by a prominent bright horizontal molding flared up and out to flow back over the hood and on into the beltline trim, which furthered the front-end emphasis. DeSoto stylists played with many grille themes, but ultimately settled for an undistinguished mix of thick and thin horizontal bars, accented by a large center emblem. Chrysler stayed with its inverted trapezoid, as for 1960-61, with different inserts to identify the various series; the proposed letter-series 300 retained the "crossbar" motif used since 1957.

The S-series Imperial was just as different from its predecessors, if less radical than the DeSoto and Chrysler. Its "face," for instance, was like the production '61 design, with a fine-texture grille set beneath a wide chromed header displaying IMPERIAL in big block letters. Flanking this were twin sets of freestanding headlamps in individual chrome pods suspended from overhanging fenders, a "Classic" throwback that still appealed to Ex. Though it sounds as ghastly as the '61 treatment, this ensemble actually had a crisp, tight look. Blade front fenders mimicked those of lesser S-series models, but rear fenders wore a single short blade terminating in bulged fender tips. On each of those was a "gunsight" taillamp, an Imperial signature since 1955. Of all the senior S-series proposals, Imperial was arguably the handsomest, DeSoto the least attractive.

Though not widely appreciated, the equally new S-series Plymouth and Dodge were supposed to be full-size cars in the traditional sense. Though early efforts looked something like the "downsized" models that appeared for '62, they were better proportioned because of longer planned wheelbases. The most intriguing of this bunch was a Plymouth "Super Sport" coupe, whose roofline combined elements of some future GM cars. A single side window, for the door (1970-91 Chevy Camaro/Pontiac Firebird), was cut up into the roof (as on the '63 Corvette Stingray coupe), and wide, reverse-slant B-pillars led to an enormous wraparound rear window that was V-shaped in plan view (forecasting the '67 Cadillac Eldorado). Though this treatment was used in watered-down form on the 1964-66 Barracuda, the stillborn Super Sport wore it much better. Other S-series Plymouth ideas seen in surviving "record" photos were a more conventional semi-fastback hardtop coupe and a back panel remarkably like that of the 1960 Pontiac.

Final S-series styling models were ready by February 1960, but the program was soon overshadowed by a scandal that shook Chrysler to its core. On April 28th, William C. Newberg, a 27-year

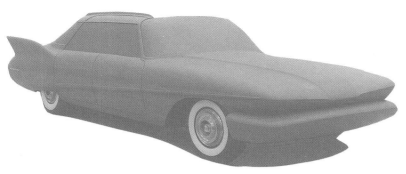

Top two rows: *These ⅜-scale clay models, shot in the Highland Park design studios in December 1957 and May 1958, represent very early work toward the '62 corporate program. Designers still clung to Chrysler's trademark tailfins at this point, and even tried "forefins" on one model (second row right). These were likely DeSoto/Chrysler/Imperial concepts. "Gullwing" fins (top row right) would surface in modified form on the '61 Imperial. Third row from top: The Chrysler/Imperial studio devised this full-size clay as one of the several proposals that lost out to the S-Series concept from Virgil Exner's Advanced Design staff. Kinked A-posts and mid-body beltline notch seem rather odd. Grille shape hints at 1960 Chrysler/ DeSoto design. Above: Another early contender for '62 was this clay from the DeSoto studio under Don Kopka. Snapped under the Highland Park styling dome in February 1959, it has a face remarkably like that of 1960-61 Mercurys—probably pure coincidence— plus the sort of "blade" front fenderlines then favored by design chief Exner.*

company veteran, was elected president; just two months later, on June 30th, he was fired by the board of directors for alleged conflict of interest: financial holdings in several Chrysler suppliers. Other officials were also dismissed, each "resignation" making big headlines.

Despite his brief presidency, Newberg was able to alter S-series plans substantially. Based on a rumor that Chevrolet would downscale its full-size Impala for '62—which proved false—he summarily ordered the approved Dodge/Plymouth wheelbase cut from 118 to 116 inches. Confusion reigned as staffers worked long hours making the original styling fit. In the process, Exner's designs lost their curved side glass and several other elements.

Though more conventional, the final '62 Plymouths and Dodges were still pretty odd—rather like overgrown Valiants. They met a poor reception—so poor, in fact, that Virgil Exner was fired, too (replaced by Elwood Engel, lured over from Ford). Meantime, L.L. "Tex" Colbert, who had engineered Chrysler's remarkable mid-Fifties comeback, briefly returned as president after the Newberg debacle. With one eye on dismal 1958-60 sales, he reviewed Newberg's plans for '62 and had second thoughts. Though it was too late to abort the shrunken Plymouth and Dodge, he *could* cancel the chunky S-series seniors and hope for the best with heavily facelifted—really "definned"—1960-61 models. Which is just what he did. At the same time, Colbert consigned DeSoto to history, concluding that a key factor in its recent poor sales was price competition from the lesser Chryslers and the grander Dodges. *Sans* fins, Chrysler and Imperial both sold somewhat better for '62. Helping Chrysler's cause that year was a new "300" line to replace the mid-range Windsor, offering the style, if not the performance, of the renowned letter-series cars at much lower cost.

Strictly speaking, the S-series was not the last DeSoto, as management briefly considered a '62 cloned from the definned Chrysler Newport. Engineering released an ornamentation drawing of it; not surprisingly, the only difference involved grille emblems. It's doubtful that Chrysler seriously intended to sell this DeSoto. Indeed, the drawing may have been done only so that officials could honestly tell the press, which had been asking pointed questions about DeSoto's future since 1959, that Chrysler really was working on new models.

On that score, one car magazine reported in early 1960 that, in light of falling sales, DeSoto would offer only a compact for '62. Presumably, this would have been a Valiant-based car instead of an all-new design, though the report—actually a bit of gossip—didn't comment on that. Perhaps they got their rumors mixed up with insider tips on Dodge's Valiant-clone '61 Lancer.

As for the stillborn S-series DeSoto, Fred Reynolds remembered inspecting the completed metal prototype six months after the program was killed. He recalled it looking grotesque, awkward, already dated. All told, he was glad the company decided not to build it, which means he couldn't have much liked the aborted Chrysler or Imperial either. Given the poor showing of the '62 Plymouth and Dodge, it's probably just as well the S-series was axed.

The upshot to this story is that DeSoto died in name only. Less than a year after its burial, the make was effectively resurrected at Dodge in the full-size Custom 880. Hastily fielded to offset underwhelming sales of those downsized '62 "standards," it arrived with a pair of hardtops, like the '61 DeSoto line, but also a hardtop wagon, four-door sedan, even a convertible. All combined a '61 big-Dodge front with a '62 Newport rear, which meant the reborn big Dodge was essentially the same car as the 1960-61 DeSoto, Chrysler and, yes, full-size Dodge. The two Custom 880 hardtops even cost about the same as the last DeSotos, and the entire line sold well, doubtless due to its conventional, inoffensive styling.

The Custom 880 continued earning good money for Dodge right on through 1966. Given that, one suspects that DeSoto's rapid decline, like Edsel's, stemmed less from a changed market than a "loser" image and increasing rivalry from sister divisions. Perhaps there's a lesson here for GM in the Nineties about the fate of Oldsmobile. We can only hope they heed it.

Chrysler management eagerly approved this full-size DeSoto convertible clay from Exner's Advanced Design staff over competing 1962 proposals from the various production studios. These photos were taken shortly afterwards, in July

'59, by which time work was underway on adapting this model's basic themes to each make in the corporate stable. Truncated rear with "chicken wing" fenders is evident, as are the thematic long hood, blade front fenderlines,

beltline kickup, and curved side glass. Triple-taillight motif revived a DeSoto hallmark from 1956-58, though in horizontal rather than vertical format.

Top two rows: *By September '59, the DeSoto studio had finalized its 1962 styling in this pair of fully finished clay models, which management approved. Changes from the earlier "theme" convertible included a more smoothly blended "chicken wing" and meeker grille insert. These are the DeSotos that would have appeared for '62 had the make not* been canceled at the end of 1960. Third row from top and above left: *The '62 Plymouth was set to wear offset hood and rear deck fins plus matching taillamps until management opted for more "equal" treatment left to right. Exner was fond of such asymmetric styling, as indicated by his wild 1960 XNR show car.* Above right: *This late mockup for* the '62 Dodge wears near final styling, *but its bright bodyside and greenhouse trim were left off, and the lower rear fenders and back bumper were reshaped before production. Also, this model was badged "Phoenix," a name not used for '62.*

The Modern Duesenbergs: Family Affairs

Legends belong to the ages and are not easily regained. That certainly applies to the mightiest of America's great Thirties Classics, the Duesenberg Model J. Pioneer automotive journalist Ken Purdy once said it "will live as long as men worship beauty and power on wheels."

Still, there are those who can't resist trying to improve on a legend, particularly when they bear the same name. That, in a nutshell, explains why the only two attempts at a modern Duesenberg—at least so far—have been made by descendants of brothers Fred and August Duesenberg, creators of the immortal J.

The first attempt began in 1964, when Augie's son Fred A. "Fritz" Duesenberg resigned as chief engine engineer for the Labeco test-equipment company to join forces with one Milo N. Record, a sales and promotion specialist lately employed at Goodyear. The impetus for their partnership was none other than Virgil Exner, the recently ousted styling chief at Chrysler. As Virgil Exner, Jr. later recounted in *Special Interest Autos* magazine: "My dad was [then] in semi-retirement. He'd done a number of designs for *Esquire* [in late 1963, interpreting] how some of the classics . . . might look in the modern era." Of Exner's four "contemporary continuations," only an updated '34 Packard went unbuilt. His modernized Mercer idea was translated into the one-off 1966 Mercer-Cobra, while his Stutz speculation led directly to the trio of Pontiac-based Stutz Blackhawk models that sold in tiny numbers from 1970 to the mid-Eighties.

But, of course, it was Exner's latterday Duesenberg that interested Fritz—and Texas real-estate baron Fred J. McManis, Jr. With dreams of raising at least $5 million in start-up funds, Fritz formed a new Duesenberg Corporation in Indianapolis, where his father and uncle had built their towering machines 30 years before. Fritz installed himself as chairman and McManis as president.

Their initial vision was a $10,000 super-luxury sedan on a 120-inch wheelbase, but that soon grew into an even costlier car with a 132-inch chassis and, briefly, an aluminum V-8 with over 500 cubic inches and 300 horsepower. Targeted yearly volume was variously quoted at between 200 and 1000 units by sources ranging from the *Wall Street Journal* to monthly "buff" magazines. Moreover, as Fritz told *Car Life*'s Ed Janicki: "We plan no annual changes [though] we might consider a change or modification after 10 years. With this price, you couldn't sell [one] and then obsolete it in two years."

After selecting a final design from 15 working sketches submitted by the Exners, Fritz okayed a prototype of what came to be called the Duesenberg Model D. Construction was entrusted to the famed Ghia works in Italy—logical, as Ghia had built most of the elder Exner's Chrysler show-car designs of the Fifties. Engineering work became a joint effort between Dale Cosper, a veteran of the original Auburn-Cord-Duesenberg concern, and

Paul Farago, fresh from birthing the Chrysler-powered Dual-Ghia. But there was never any rush to completion, because financing was slow and hard to come by. In fact, hopes of attracting new money prompted the prototype's first public showing, which didn't come until the spring of 1966.

Like its hallowed forebears, the new Model D had grand proportions: It was a four-door brougham sedan measuring 137.5 inches between wheel centers and 245 inches overall—half an inch longer than a contemporary Cadillac limousine. The announced price was a lofty $19,500, but included automatic transmission (Chrysler TorqueFlite), automatic climate control, all-disc brakes (big Airheart units), torsion-bar front suspension, chrome wire wheels, and power everything. Per Duesenberg tradition, back seaters could scrutinize their own speedometer and clock; they also enjoyed a separate radio, fold-out writing tables, even a TV and bar. Interior trim was top-grade leather with solid mahogany accents. The exterior blended nostalgic elements—razor-edge roof, center-opening doors, clamshell-shaped wheel openings—with trendy stuff like hidden headlamps. With 350 bhp from a stock 440 Chrysler V-8 (the 426 Hemi was considered but rejected, as was all-independent suspension), the Model D had good performance for a 5700-pound biggie.

But this first modern Duesenberg never went any further. Though plans were afoot for limousine and four-door convertible models, simple start-up of sedan production demanded $2.5 million, and the money was nowhere to be found. So, after a few months in the limelight, Duesenberg Corporation faded away. Which was a real shame. According to the few who've driven it, the Model D handled well for its size and had all the luxury anyone could want. But the potential demand for such a costly "retro" car in 1966 was tiny if not non-existent, and the concept itself was probably flawed. As *Car and Driver* later opined, the Model D seemed the "perfect 1934 dream car. . . . [Fred and Augie] would have kept up with the times."

Even less ambitious was the revival attempt by Fred and Augie's grand nephews, Harlan and Kenneth Duesenberg, which surfaced in 1976. After forming a new Duesenberg Brothers Company, the pair hired Robert Peterson of Chicago's famed Lehmann-Peterson limousine works to engineer another "modern Duesenberg." The result appeared three years later as little more than a customized Cadillac. Power came from a stock 425-cid fuel-injected V-8 with 195 bhp (SAE net). The chassis was Cadillac's too, though with a unique 133-inch wheelbase halfway between that of the then-current DeVille sedan and Fleetwood limousine.

Though its intent was laudable, this car was a very pale reminder of Model J glories. At least price kept with Duesenberg tradition: an astronomical $100,000. But styling was boxy, slab-sided, heavy-handed yet bland. In front, for example, were

Top: *First shown in 1966, the Model D was the first attempt at a postwar Duesenberg with basic design links to the hallowed Classic-era Model J. Styling stemmed from a 1963 "contemporary continuation" concept sketched by ex-Chrysler design chief Virgil Exner,* assisted by son Virgil, Jr. Classic "cues" like a big radiator-type grille and "clamshell" fender shapes were deliberately melded with hidden headlamps and other trendy touches. Above: Built by Ghia of Italy, the lone Model D prototype was fully functional *and beautifully finished inside and out. Spacious cabin was awash in fine leather, broadcloth, and wood trim (left), and preserved a Model J tradition with full instrumentation that included a stopwatch and altimeter (right).*

stacked quad headlamps outboard of hidden driving lights astride a squat square grille that could have come off a late-Seventies Lincoln. The capper was a garish "bow-tie" front bumper that looked like a bushy moustache.

Harlan and Kenneth planned to build their Duesenberg one order at a time, selling direct from a small plant in the Chicago suburb of Mundelein (the intended location was later changed to Evanston, Illinois). But again, funds ran out after a single prototype was built. That car survives today, as does the Model D.

Ironically, the "1980" Duesenberg was partly motivated by the brothers' desire to make up for the stillborn '66 revival, thus restoring luster to the family name. Perhaps some future member of the clan will finally succeed in producing a modern Duesenberg with the spirit and excellence of the great Model J.

Top left: *Though one magazine chided it as "the perfect 1934 dream car," the Model D was handsome for the late Sixties and arguably still looks good. The large circle at the base of the grille was meant to recall the old-fashioned crankhandle hole of many early-Thirties* Classics. Top right: *Preserved today at the Auburn-Cord-Duesenberg Museum in Auburn, Indiana, the Model D "brougham sedan" would have been joined by limousine and convertible sedan models had production materialized, but funds ran out soon after* the one prototype was built. Above: *The Model D shows a clean rear aspect, emphasized by modest taillamps housed neatly within the bumper, plus a small "formal" backlight.*

The second stab at a modern Duesenberg was this Cadillac-based sedan begun in 1976, though not widely promoted until 1979-80. Production seemed imminent by then, but was soon derailed by a second energy crisis and a sharp new recession. Blocky styling (top and above right) contrived to hide the Cadillac origins and gestured to genuine Duesy hallmarks with eagle insignia and "bow-tie" front bumper. Interior (above left) was lush but rather ordinary, which belied the announced $100,000 price. The car was planned to be built one order at a time in a small facility in suburban Chicago.

Ending Edsel: Hopes Before the Fall

You can't ignore lead times if you want to understand why some cars succeed and others fail. Consider Edsel, arguably the biggest automotive flop of all time. As you might know, it was conceived in heady 1953-54 by a Ford Motor Company then bounding back strongly from near collapse in the late Forties. Led by board chairman Ernest R. Breech, optimistic Dearborn managers laid expansionist plans to match General Motors model-for-model with a GM-style five-make hierarchy involving a separate new Continental Division and a second middle-class brand to bolster Mercury. The latter made appealing sense at a time when the medium-price field was booming. In record-setting 1955, for example, Pontiac, Buick, and Dodge built nearly two million cars combined.

But with the three-year development cycles then customary in Detroit, Edsel didn't arrive until late 1957—by which time the entire market was depressed and the medium segment had shriveled from 25 to some 18 percent. Though hoping to sell 100,000 of its debut '58 models, Edsel Division built only a little over 63,000, which was actually fair going for a new line in a recession year. But from there it was all downhill. After just under 45,000 for '59 and a mere 3000 for 1960, Edsel was canned on November 18, 1959 as a colossal blunder and object lesson in the limitations of "motivational research."

Edsel's here-today gone-tomorrow existence might imply that Ford gave up too easily, but it isn't so. Before the end, Dearborn planners contemplated many ideas for 1959-60, one of which was actually a great success.

Though first envisioned as a more expensive and powerful "super Mercury," Edsel ended up positioned between Ford and Mercury. The initial 1958 offerings comprised Ranger and step-up Pacer series sharing chassis, running gear, and bodyshells with the 1957-58 Fords, and a smaller group of senior Corsair and Citation models bearing like relationships to contemporary Mercurys. As we know, the '59s were reduced to Ford-based Rangers and Corsairs plus Villager wagons, but the original plan had been to continue the broad '58 lineup, with seniors again owing much to Mercury.

The stillborn big Edsels would have been quite different from their '58 forebears because of the all-new platform decreed for that year's 20th anniversary Mercurys. Typical of the times, it was longer, lower, and wider, but also much cleaner than the 1957-58 design. It was more practical too, with a space-making, slimmed-down dash and huge new expanses of glass for outstanding visibility. Naturally, the '59 Merc-based Edsels would have shared these features, as well as a wide-track "cowbelly" chassis. Wheelbase would have been the same 126 inches used for all '59 Mercs save the opulent Park Lane (128), up two from the '58 Citation/Corsair.

To set all their '59s apart, Edsel stylists toyed with variations on established themes. Surviving photos show the stillborn seniors combining Merc greenhouse with distinct Edsel faces. Designers were evidently satisfied early on with lowered head-lights and a blunted version of the infamous "horse-collar" grille, for both survived to the Ford-based '59s.

But the big '59 Edsels were ultimately dropped, though not until fairly late in the game. The reason, of course, was dismal sales of the '58 models—barely a third of the model-year total, which itself was a mighty disappointment to Dearborn.

The 1960 Edsel evolved as nothing but a restyled version of that year's bigger, all-new standard Ford. As later recorded by Edsel PR director C. Gayle Warnock in his book, *The Edsel Affair*, this decision had come way back in April 1958, when it was abundantly clear that the Edsel experiment had gone awry. The man behind it was none other than the no-nonsense Robert S. MacNamara, then a Dearborn vice-president, who declared the 1960 Edsel should be merely "a variation on the Ford car, using the same major components with modified front and rear ornamentation."

Even so, "instant recognition" was still deemed important to Edsel sales when work on the 1960 began. *Special Interest Autos* magazine suggested as much in 1970 with a rescued photo of the front-end treatment originally intended. This was somewhat like the final production design save a prominent bright central bar running up from the bumper into a chrome-edged nacelle. Both bar and nacelle were roughly triangular, with the latter blended smoothly into flanking cross-hatched sub-grilles. The hood formed its top portion, and continued its line rearward in Edsel's customary tapering-vee bulge. If not exactly timeless—the effect reminds one of TV's Ollie the Dragon—the treatment was at least identifiably Edsel and far less prone to joke-making than the original horse-collar.

Trouble was, it meant unique hood and grille stampings, and that was too much for the sales-conscious MacNamara. His cost concerns also overruled 1960 Corsair models (which would have worn a wide, tapering swath of brushed metal on their lower flanks), as well as a rear-end treatment that used Ford's new flat-fin rear fenders to revive a '58 Edsel hallmark: "gullwing" tail-lights. The 1960 Edsels thus bowed in October 1959 with just five Rangers and two Villager wagons bearing a split grille remarkably like that of the previous year's Pontiac, plus four vertical ovals stuck awkwardly onto the Ford rump for tail- and backup lights. There were also bullet-style parking-lamp housings and big E-D-S-E-L lettering on the lower rear fenders, but most everything else was 1960 Ford. Not that it mattered much, for Edsel's plug was pulled barely a month after the '60s went on sale.

Yet that still wasn't quite the end of things. Even before the

The Edsel was initially conceived in 1953-54, and this early sketch (top left) has a certain flavor of the '56 Lincoln—note the rear-quarter windows, hooded headlights, and the shape of the wheel openings. The vertical grille theme is already apparent, however. Serious proposals for the '58 Edsel (top right and center) all carry slightly different versions of the infamous "Oldsmobile sucking a lemon" grille, but note that only one incorporates dual headlights, and only on one side (top right). Early

proposals for the '58 used single headlights, likely because duals weren't yet legalized. The complicated rear-quarter sculpturing on a clay model shown at a July 19, 1955, presentation (center left) was thankfully rejected. An August 17, 1955, clay (center right) shows a car very close to production form, though side trim is missing and the rear bumper design would be changed before production started. For 1959, the Edsel shared only the Ford bodyshell, but there had originally been plans for a

Mercury-based 1959-60 Edsel, just as there had been in 1958. The two proposals shown (bottom row), which clearly have the '59 Mercury greenhouse, would have been the 1959 top-line models. As in 1958, all Edsels would have shared identifying styling details, such as grille, taillights, and bumpers. The car on the right sports side trim that looks like an opened-up version of what actually appeared on the '59 Rangers.

instant success of its new 1960 Falcon compact, Ford was working on a slightly larger, plusher version, and for a time this was planned for Edsel. Whether it would have replaced or merely supplemented the standard-size models is unclear. What is clear is the rationale behind it: Edsel was going nowhere along with other medium-price cars while smaller economy models were selling like crazy, so perhaps Edsel might still have a future with a compact.

Sound thinking except for one thing. Though "Edsel" honored the only son of old Henry Ford and the artistic force behind great Classics including the first Lincoln Continental, few people had really liked the name—and they liked it even less once Edsels earned their reputation as poorly built, gimmick-laden, slow-selling gas guzzlers. Mercury, on the other hand, came through the recession battered but with its honor intact, so it made far more marketing sense to put that name on the new "grownup" compact. Which is how the erstwhile small Edsel came to bow on March 17, 1960 as the Mercury Comet. Its one real styling change was a full-width two-tier grille instead of a modest vertical schnoz with horizontal sub-grilles. Even the slanted oval "cat's eye" taillamps were retained—a visual remnant of Edsel's 1960 standards.

How Comet would have fared as an Edsel is anyone's guess, but it was a huge success as a "Mercury" (though it didn't bear that nameplate until '61). Despite an abbreviated season, Comet racked up 116,331 sales in its debut 1960 model year, about 5000 more than Edsel had managed over three years.

That should put the lie to the old saw about Edsel losing some $250 million, which was actually what it had cost just to launch the make and was more than offset by high early Comet sales. Then too, Ford wouldn't have been able to build so many Comets or Falcons had it not been for Edsel plants, which were quickly retooled once Edsel was dropped to the benefit of compact sales.

There's an interesting postscript involving Edsel's intended 1960 styling. Though Ford built at least one full-size mockup, a Corsair convertible, it was destroyed per company practice, leaving only record photos to confirm its existence. But one Ohio collector saw those pictures and decided to replicate what Ford had planned, if only for history's sake. He did it by restoring a 1960 Ranger hardtop coupe as a never-was Corsair complete with "dragon's tooth" front and brushed-metal side trim, all hand-fabricated, of course. But he could do nothing with the back, as photos of the rejected rear-end styling have yet to surface. Even so, his project won a People's Choice Award at one Edsel Owner's Club meet in the late Eighties, and we picture it here to show what might have been.

As for the entire Edsel experiment, historians agree that it failed for two reasons. The main one was plain old bad timing. As Art Railton of *Popular Mechanics* said at the time, "Edsel was born too late." *SIA*'s Michael Lamm later echoed this view, saying the car's "aim was right, but the target moved." Ford's cynicism also played a part. Because it was "researched to death," as Lamm observed, the Edsel was a triumph of marketing style over automotive substance, and was thus almost designed to fail.

Yet one wonders whether Edsel might still be around had it arrived three years either side of 1958 *and* been truly different. One long-overlooked factor in Edsel's quick demise is the product's unintended "overselling" a good two years before it appeared. This came mostly through "rumor mill" reports that seemed to promise all manner of exotic features, thus setting public expectations far above what any Detroit maker could have met at the time.

Ultimately, the Edsel was not just too late but too little—not a "car of tomorrow" but just another "car of today." If no worse than its rivals, it was certainly no better, so its rejection was perhaps a foregone conclusion. After all, nobody likes broken promises.

The poor showing of the '58 Edsel in the recession of 1958—and indeed that of most middle-priced cars—caused Ford top brass to reconsider the role of its new car. As a result, model offerings were cut drastically from 18 in 1958 to 10 in 1959, and the base price now topped out at $3072, as compared to $3801 in 1958. The result was perhaps predictable—production dropped by nearly 30 percent in 1959 compared to an already disappointing 1958. What then to do for 1960? Product planners already had a

two-series lineup of Edsels in the pipeline: Ranger and Corsair (plus two Villager station wagons). However, this was not to be, as only a five-model Ranger series made it to market (plus two Villagers). The Corsair series was to have included a hardtop coupe (top left) and a convertible (above left). The latter, sporting the spinner hubcaps that were optional in 1958 and '59, is seen here as a prototype. Corsairs would have been the top-line models, as they were in 1959, featuring higher-level trim than the

Rangers that did make it into production. On the exterior, two chrome spears enclosing a brushed anodized aluminum insert were planned, compared to a single strip on the Ranger. No prototype now exists. But Charles O. Wells, of Medina, Ohio, liked the styling of the Corsair, so when he did a body-off restoration of his '60 Ranger hardtop (which he bought from a friend who had found it in an Oklahoma junkyard), he underwent the expense of recreating the prototype (above right).

At one point in the design process, the '60 Edsel was to have strutted a vertical-bar grille. It would have differed considerably in detail from the 1958 and '59 Edsel grilles in that the vertical portion would have been a single narrow "tooth" flanked by eggcrate grille assemblies incorporating quad headlights. However, Ford executive Robert S. McNamara had proposed in April 1958 that the 1960 Edsel should become simply "a variation of the Ford car, using the same major components with modified front and rear ornamentation." The '60

Edsel thus emerged as a thinly disguised Ford wearing a poor copy of a 1959 Pontiac grille, unique parking lights and side trim, and upright taillights, replacing Ford's "setting sun" units. When Charles O. Wells decided to restore his '60 Edsel, he wanted to reproduce the unique grille and trim that were seen on an early 1960 Edsel prototype. He blew up photos of that original styling proposal in order to make wood forms and molds. The grille assembly (top) was cast in aluminum, the egg-crate section from rolled steel bar stock. The front bumper

was created from two pieces welded together. On the side (above), the front of the chrome spear came from a '59 Edsel (shortened to fit), as did the Corsair script. Today, who would know that this Cadet Blue Metallic and Polar White Edsel isn't the real prototype? No matter, it won a People's Choice Award at the Edsel Owners Club meet in Dearborn. More important, this is the only '60 Edsel that tells us what might have been.

Sports-Car Memories: Fabulous Forgotten Fords

Top: *Tucker designer Alex Tremulis conceived the wildly aerodynamic Thunderbird Mexico in 1956 to further the budding competition career of Ford's "personal" car. Named in hopes that the famed Mexican Road Race might be revived, it was shaped for zero rear lift and a 0.22 drag coefficient—quite low even now. A smooth underbody created a stabilizing "boundary air layer" that was* channeled back to exit between prominent tailfins. With all this, Tremulis claimed the Mexico capable of a safe 200 mph on just 240 horsepower. But the T-Bird was never meant to be a track star, and the Mexico never progressed beyond this ⅜-scale model. Above left: *One of numerous full-size mockups leading toward the first "1964 ½" Mustang, a convertible in the "Allegro" series of* proposals circa 1962. Above right: *First shown in late 1963, the Mustang II provided a mildly exaggerated preview of Ford's forthcoming sporty compact "ponycar." Differences from production involved a lower roofline, a somewhat bulkier rear deck, a pointier front, and no bumpers.*

Top: *Mustang I was a very different colored horse from Ford's production ponycar. Completed in 1962, it rode a trim 99.2-inch wheelbase, had swoopy fiberglass bodywork, and carried a 1.5-liter V-4 behind its two-seat cockpit to drive the rear wheels through a four-speed manual gearbox. The drivetrain* stemmed from Dearborn's aborted *"Cardinal" small-car project that was transferred to Ford Germany for the Taunus 12/15M. Though briefly considered for production as the sporty compact desired by Ford Division chief Lee Iacocca, Mustang I was rejected as too costly for the targeted $2500 retail* price—*and "too far out" for likely high-volume sales.* Above: *Ford built two versions of the Mustang I: a non-running mockup and a fully driveable model. Here, Ford engineering vice-president Herb Misch gets behind the wheel of the "runner" as styling V-P Gene Bordinat looks on.*

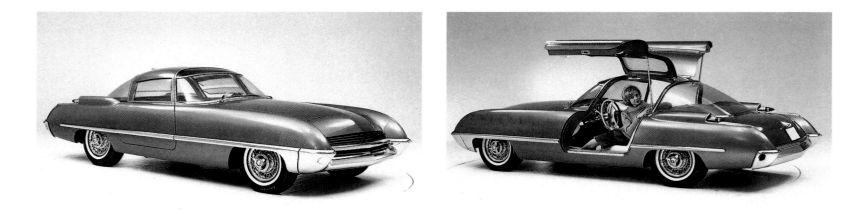

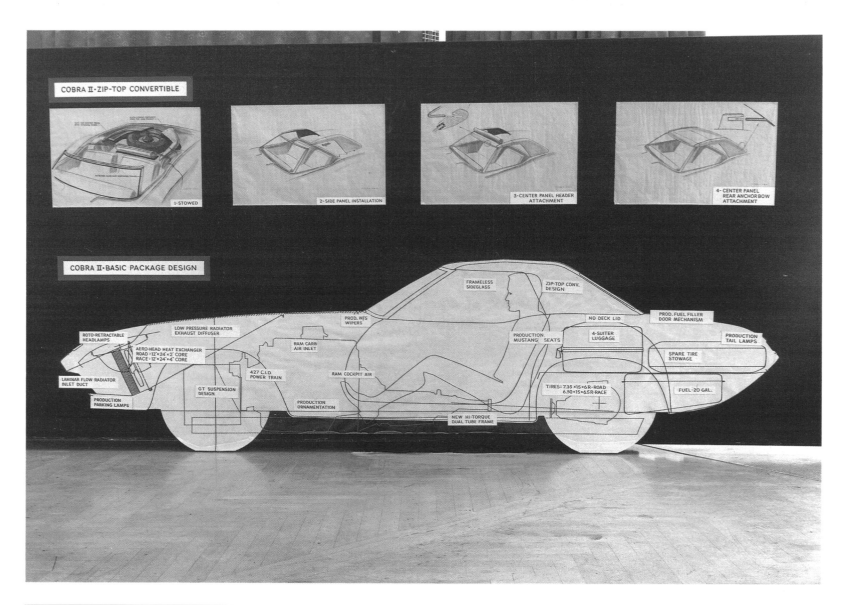

Top left and right: *Five years before Mercury's Cougar, Ford showed this jazzy closed two-seater of the same name. Unveiled at the 1962 Chicago Auto Show, it carried a big-block 406, then Ford's largest V-8, plus electrically operated gullwing doors a la the mid-Fifties Mercedes 300SL coupe. "Spacey" styling was typical of period Detroit* showpieces. *Wheelbase was a tidy 102 inches, overall length 180 inches, height just 49.5 inches. Some magazines portrayed the Cougar as a new showroom luxury Ford in the spirit of the original Thunderbird, but it was not to be.* **Above:** *This detailed elevation shows the people/powertrain package worked out for the Cobra II convertible circa 1965,* another of the many sports-car notions Ford toyed with long after building its final two-seat T-Bird. Designers planned for a big-block 427 V-8 as well as a novel multi-position fabric "Zip-Top" (worked out in the illustrations above). Though the design looks production-ready, it's doubtful Ford ever entertained serious thoughts of building this car for sale.

Top: *First seen in 1963, the Cougar II was built on the 90-inch-wheelbase chassis of the contemporary AC/Shelby-Cobra to juice up Ford's "Total Performance" image. The car survives today in like-new shape, and remains eye-pleasing. Powered by Ford's new big-block 427 V-8, it was rumored as an imminent reply to Chevy's new Corvette Sting Ray. But though Cougar II reportedly came very close to production, it was nixed by Dearborn bean-counters for the usual reason of "low volume equals low return." Alleged top speed was in the region of 170 mph. Other photos: Cougar II later gained this companion convertible called Cobra II, though it's also been known as the "Bordinat Cobra" after Ford design chief Gene Bordinat, one of its prime advocates. Again, Ford looked ready to attempt a modern high-performance Corvette challenger, but there was really no need, especially once Mustang joined the Shelby Cobra itself in Dearborn's sporty-car stable. Too bad—top up or top down, this car looked great.*

Two Dearborn delights from the days when mid-engine design seemed the wave of the future for production sports cars. Top and second from top: It's a DeTomaso Mangusta all right, but shot in the Ford Design viewing court circa 1968 wearing "Shelby Mark V" i.d. This suggests Ford briefly thought of selling the Italo-American hybrid (which bowed in '67) as a Stateside replacement for the open Shelby Cobra—no problem given

the all-Ford mechanicals. Unfortunately, the willowy, cramped, and tricky-to-handle Mangusta was not terribly saleable. Ford's better idea here was to get DeTomaso working on the more practical 1970 Pantera. Other photos: 1971's Mach 2 show car hinted that the next Mustang generation might be a midships design—which, of course, it wasn't. Still, this dazzling follow-up to 1968's conventional Mach 1 (which previewed

production '71 'Stangs) was a nice reply to numerous mid-engine "dream" Corvettes. Styling blended American muscle car themes with the tailored look of Ford's mid-Sixties GT40 and Mark II endurance racers. Note the "tunnelback" rear window, "limousine" door cuts, twin racing-type fuel fillers (on the B-posts), and one of the first uses of "window-in-window" door glass.

Looking taut and purposeful, the mid-Eighties Cobra 230 ME was submitted by Ford International in Europe as one of three prospects for a $30,000 midships sports car to be built in limited numbers as a counter to Chevy's latest Corvette. Other contenders for the project, which was launched in 1982 with the code name GN34, were a Dearborn mockup and the running "Maya" prototype from Giorgio Giugiaro's Ital Design. Despite its "230" tag, the ME carried a 2.5-liter version of the Mustang SVO's 2.3-liter intercooled turbo-four, sited transversely behind a two-seat cockpit; 16 valves and a hotter cam reportedly boosted horsepower to 300. Production was briefly "go," and assigned for economic reasons to the small Chausson works in France. But that dubious decision combined with a weakening dollar, development delays, and spiraling costs forced cancellation of the project by late 1986. Had it been built as designed, the 230 ME would have bowed for 1989 or '90 with a 95.6-inch wheelbase, all-independent suspension, four-wheel anti-lock disc brakes, sub-3000-pound curb weight, and estimated 0-60-mph performance of six seconds or less with five-speed manual transaxle.

Dearborn Discards: The Fords in Nobody's Future

The wood-bodied Sportsman convertible added a much-needed touch of glamour to the warmed-over prewar Fords of 1946-48, but high price meant low sales. Still, Ford considered continuing the Sportsman as one of its all-new '49s. In fact, it considered two: a new-design convertible (top) and the first-ever Sportsman club coupe (above left). Unlike the earlier ragtop, the wood here was decorative, not structural—a modest framework applied to the bodysides alone. Yet even this was deemed unnecessary for the most changed Ford since the Model A, so no '49 Sportsman was built beyond these mockups. A more serious question was whether to retain fastback styling for '49. Ford obviously gave it some thought by building at least this one full-size mockup (above right), then had second thoughts. It was just as well, for fastbacks quickly faded from public favor in the early Fifties.

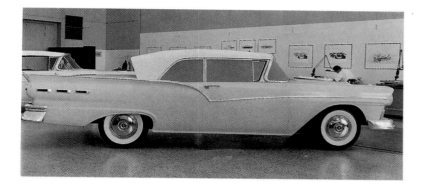

The 1957 Ford was among the most ambitious new-model programs in Dearborn history, and no expense was spared exploring design ideas and even body types. Top left and right: A double-sided full-scale clay from 1956 for "Parklane" wagons. Note the slanted C-posts and curved D-pillar/tailgate, plus the two-door's high-cut windshield and the four-door's A-pillar air scoop (as on the '57 Mercury Turnpike Cruiser). Second row: Close to final '57 Ford styling, but triple rear-fender vents didn't make it. Neither did a convertible sedan (right) nor a convertible coupe (left) with '58 Lincoln-style aft top styling. Third row: Odd overhanging rear roofline was long favored, but didn't survive. High windshield with flanking scoops (left) was later lowered (right), when C-pillars added tacky extra trim. Above: This elevation suggests that the singular "bubbletop" Crown Victoria pillared hardtop might have returned from 1955-56, but accountants surely vetoed a '57 version due to very low Crown Vic sales.

Top row: *Two of many workouts for the Thunderbird's 1957 facelift. Jazzy front wheel-arch sculpturing (left) was axed in mid-1955 when the new '56 Corvette appeared with something similar. Full hardtop side windows would have been a welcome aid to visibility. Fins were favored from the first for '57 (right), but designers debated the rest of the tail. This model's right side wears the basic* treatment ultimately chosen over gimmick ideas like that on the left. Second row: *A pair of restyles proposed for the '58 standard Ford. Early tryout (left) shows more preoccupation with heavy rooflines and busy grilles; later attempts (right) were more restrained, with a focus on lending a family resemblance to the evolving '58 T-Bird.* Third from top: *We should perhaps be* glad this hideous creature was left stillborn. It likely aimed at '58, though compound-curve windshield hints at '59. Above: *The '59 Ford was close to final when this hardtop clay was photographed. Gracefully airy rear roofline wasn't used, but may have prompted the even prettier and sleeker 1960 Starliner.*

Top: *An early mockup for 1959 shows how designers strained to give standard Fords a '58 T-Bird-type face with "gullwing" headlamps and a big bumper/grille. This hardtop-wagon clay also envisioned fancifully huge glass areas, throwback center-opening doors, and spacey lower-rear-fender "canards." Though Ford never offered hardtop* wagons, Mercury did for 1957-60. Other photos: *Three views of the most radical big-Ford design in final contention for '59. Reverse-slant rear window (second row) was borrowed from the '58 Continental Mark III and would likely have lowered electrically as an option. The idea would return on the 1963-64 "Breezeway" Mercurys. "Gullwing" rear* windows seem to be borrowed from Edsel, *while the circular taillamp and parking-light housings were evidently made big enough to double as bumpers. Rakish "checkmark" side trim was still a favored Ford motif when this model was shot (likely in early '57). We should probably be glad this one got away, too.*

Top: *This early effort toward the third-generation '61 Thunderbird, undoubtedly from Bill Boyer's studio, combines 1958-60 lower-body themes with a slantback roof and an obvious jet-plane tail. Boyer favored missile and aircraft themes for their presumed youthfulness.* Second row, left: *The '61 Bird's eventual "fleet submarine bow" and big "flower pot"* taillights are evident here, but this model's fin and roof treatment were vastly—and tastefully—toned down. Second row, right: *A rather heavy-looking mockup, again from Boyer, with a wild rear evidently inspired by Ford's "Gyron" experimental, which was shown in 1961.* Third row: *A later variation on the Gyron theme (left) makes the* "missile" mockup (right) seem almost conservative. The latter's lower body is similar to the final '61 design, but not its back panel or the improbable canopy-style roof. Above: *Central decklid fin seems very out of place on this otherwise formal-looking model. Note the A-pillar treatment here as well as the gaudy "electric shaver" back panel.*

Top and second row: *Shot in the studio with a production Mustang and by itself outdoors on a snowy Dearborn day in December 1964, this minicar, badged "Colt," was a likely follow-up to 1963's "Hummingbird" project, which envisioned a "world car" long before the American Escort. Though Hummingbird was planned around a small twin, the* Colt's *long hood suggested a considerably larger engine. Note, too, the truncated tail, which predates the similar AMC Gremlin treatment by a good five years.* Other photos: *Believe it or not, this big bruiser was mocked up around 1969 for what emerged five years later as the subcompact Mustang II. Early efforts like this obviously didn't foresee the rapid decline in demand for traditional performance cars then already evident— nor the first energy crisis of 1973-74. The turn to a "smaller is better" Mustang was prompted in part by the early-Seventies popularity of sporty import coupes like Ford's own European Capri.*

Ford's Forgotten Carousel: Missing the Modern Minivan

Chrysler Corporation likes to boast that its 1984 Dodge Caravan/Plymouth Voyager pioneered the minivan concept, but that's only true of front-wheel-drive designs. The real originator was Volkswagen, which had sold rear-drive minivans since the early Fifties in the Beetle-based Type 2 Microbus (through 1979) and the later Vanagon. There was also Chevrolet, which had imitated VW for 1961 with a trio of trucks derived from its compact Corvair, another air-cooled rear-engine car. Besides a novel pickup called Rampside, Chevy offered Corvan commercial and Greenbrier passenger vans remarkably similar in size and shape to the Chrysler models of nearly a quarter-century later.

That same year, 1961, Ford trotted out a group of Falcon-based compact trucks called Econoline: initially a pickup, commercial van, and windowed passenger van. All had conventional rear drive and boxy "cab-over-engine" bodywork like the VWs and Chevys, but hewed to orthodoxy with water-cooled six-cylinder engines mounted in front. Simple and reliable, Econoline sold well from the first, easily outpacing the more trouble-prone Corvan/Greenbrier. It also outlasted them in the marketplace. Almost predictably, though, Econoline grew quite a bit larger when first redesigned for 1968.

Then again, the truck market itself was growing—and subdividing, just as automotive classes had done with compacts, intermediates, and personal-luxury models. When buyers began clamoring for utility vehicles in the image of the wartime Jeep, Ford obliged them with the Bronco for 1966. Meanwhile, custom vans had become the rage with many younger people of the "hippie" stripe, and commercial users were demanding bigger, heftier vans as well as pickups. That set Ford to working on an even larger Econoline. Developed as project "Nantucket," this third-generation design was pretty much locked up by 1972. Introduction was slated for model-year '75.

But something else was afoot, or so thought one Lido Anthony Iacocca, the former super-salesman from Pennsylvania who had fast climbed the ladder at Ford Motor Company to become its president in 1970. That presidency was partly his reward for "fathering" the original Mustang ponycar and some newer Dearborn winners. Of course, Iacocca wasn't solely responsible for the Mustang's huge success, as he (and others) liked to claim, but he could spot emerging market trends.

With vans bigger than ever, in popularity as well as size, Iacocca began wondering whether there might be a market for something smaller and less "trucky"—in other words, a more maneuverable "car-like" van. Suitably priced, it could have great appeal to "vannies" and smaller families for camping and other recreational uses, as well as light-duty urban commercial. For some buyers, this "minivan" might even be a roomier, more versatile alternative to the traditional station wagon.

Accordingly, Iacocca approved development of what came to be called the Carousel. Conceived as a more compact "garageable van," it was planned around dimensions that harked back to the first Econoline, plus, at first, an all-new drivetrain. But cost concerns soon prompted changing the concept to essentially a cut-down version of the just-completed Nantucket, with most of its innards but different outer sheetmetal and more carlike amenities.

The revised approach offered benefits apart from reducing development time and tooling costs, not to mention retail price. Cutting down Nantucket's inner body and ladder-type frame promised a very strong, rigid platform; the separate chassis, combined with a heavily insulated floorpan and thick bushings beneath driveline and body, promised a quiet vehicle without the unpleasant rough-road "drumming" that marred so many unit-construction cars. Of course, front drive never figured in the equation. At the time, the configuration was not yet fashionable for Detroit cars, let alone trucks. Even today it's sometimes deemed undesirable compared to rear drive because it tends to limit towing ability, an important consideration for most truck buyers.

Carousel styling ended up being largely the work of designer Dick Nesbitt, who was also a principal in shaping the 1974 Mustang II. Striving to avoid the "school bus look" typical of contemporary big vans, he penned a glassy roofline not unlike that of Chevy's classic mid-Fifties Nomad wagon, with slanted B-posts, thin C-pillars, and three windows per side: one for each front door and two behind. The rearmost panes curved slightly around the back to meet modest D-posts, matched by wrapped, tri-color vertical taillamps. Ribbed appliqués in a dark charcoal color adorned lower body and tailgate to respectively lower and widen appearance, though they would have been omitted on the woody-look "Squire" version that Ford marketers wanted. The tailgate itself was a drop-down wagon-style affair with retracting window.

Some Carousel sketches furthered the Nomad look with a slight roof-panel "step-down" behind the B-posts; etched with slim longitudinal lines, it was the natural spot for an optional luggage rack. Up-front ideas involved variations on Ford's then-favored "power dome" hood theme, with four round headlights in separate square bezels astride big, mostly rectangular grilles.

Like Nantucket but not earlier Econolines, the Carousel had a definite, if stubby nose. This allowed the engine to sit farther forward, which opened up extra space and lessened noise inside. Iacocca also wanted the protruding front sheetmetal as a "crumple zone" for absorbing energy in collisions—a sales plus for the safety-minded (including, one presumes, the likes of Ralph Nader).

As was becoming *de rigeur* for vans, a sliding right-rear door provided access to Carousel's rear compartment, for which there was no shortage of ideas. Nesbitt later recalled that one of the

Although Chrysler Corporation gets credit for marketing the first "garagable" van, Ford Motor Company, with the encouragement of then-President Lee Iacocca, actually had one under development in the early Seventies. The one seen here (top) was sketched in October 1972 by Dick Nesbitt, who was a designer for Ford at the time. Although based on the standard Econoline van, proposals featured all-new sheetmetal, unique glass treatment, a lowered roofline, and a sliding side door for easy access to the rear seats. Planned introduction was for the 1975 model year. It was Iacocca and Hal Sperlich, product planning head at Ford, who decided early on that some form of a van-type layout would offer the most design flexibility and versatility for an optimum family vehicle. At the time, this meant it would have to be truck-based, rear-wheel-drive vehicle. One of the keys to developing a versatile, family-oriented van was a low overall height. This would not only allow for comfortable entry/exit, it would also mean that it could be parked in the average garage without scraping its roof on the way in. Among Nesbitt's proposals for the garagable van's front end, done in November 1972, were two (above, left and right) that focused on modified interpretations of Ford's "power dome" hood and variations of the upright, formal grilles then in vogue at Ford (and elsewhere in Detroit, for that matter).

more striking proposals was a U-shaped "lounge" formed by a rear-facing bench behind the front cabin and an inward-facing bench along each sidewall. Individual "captain's chairs" seated driver and front passenger.

After weeks of what Nesbitt described as "frantic effort," the Carousel was ready for management review in November 1972: a full-size clay with Squire-type side trim and conventional front-facing second and third bench seats. Iacocca, product-planning whiz Hal Sperlich, and even chairman Henry Ford II were all impressed enough to approve construction of a running metal-bodied prototype for production engineering purposes.

For a time, everything was go. Ford even planned to launch Carousel alongside the new '75 Econoline. But when OPEC turned off America's oil tap in late 1973, gas supplies dwindled, prices soared, and people lined up to buy small economy cars, shunning bigger vehicles as never before—cars and trucks alike. With that, Ford decided the Carousel had no future, and abruptly canceled it.

What happened next is the stuff of legends. Iacocca, fired by HFII in 1978, took Sperlich along to rescue faltering Chrysler, then sealed its recovery by reviving the Carousel concept for the T115 Caravan/Voyager, thus uncovering the huge minivan market that Chrysler still dominates today. Ironically, the T115 was much like what the Carousel would have been except for being based on

a front-drive passenger car (Chrysler's life-saving '81 K-body compact).

In retrospect, then, Ford was unwise to throw out its baby-van idea with the energy-crisis bathwater. Though RV sales remained sluggish, the overall market was again healthy when the '75 Econoline appeared. Considering how Chrysler later fared with minivans, it's likely Carousel would have been another sales blockbuster *a la* Mustang—and another big feather in Iacocca's fedora. While it probably wouldn't have prevented his firing, it would have given Dearborn a big competitive edge in the light-truck field for the *second* energy crisis that hit in 1979.

And there's the greatest irony of all, for the Carousel promised to be far more fuel-efficient than any full-size van would be for a long time to come. Though its likely powertrains remain unclear, lower weight implied a nice performance/economy balance with a modest six, and better than adequate performance with Ford's 302 small-block V-8, which certainly would have fit.

But Dearborn missed its chance to define the modern mini-van and has had to play catch-up ever since—first with the trucky rear-drive Aerostar, then the Chrysler-like Mercury Villager, which bowed no less than 20 years after Carousel's intended debut. If there's a moral to this story, it is that a good idea remains a good idea no matter who takes advantage of it first.

Though it may look pretty "square" in today's aero-design age, the finished clay model of the Carousel (above) actually lost much of the boxy look of the Econoline Club Wagon on which it was based. First off, it was two-and-a-half inches lower and less truck-like in

appearance. Note, for example, the slanted B-pillars, which give it more of a passenger-car look. Ditto the D-pillar area at the rear, where the side windows wrap around for better visibility. Instead of the double rear doors found on the workaday Econoline van, the Carousel

would have used a station wagon-style retractable, self-cleaning electric tailgate window. This version of the Carousel, dated November 8, 1972, strutted woodgrain bodyside trim, another feature designed to make it seem more like Ford's car-based Country Squire station wagon.

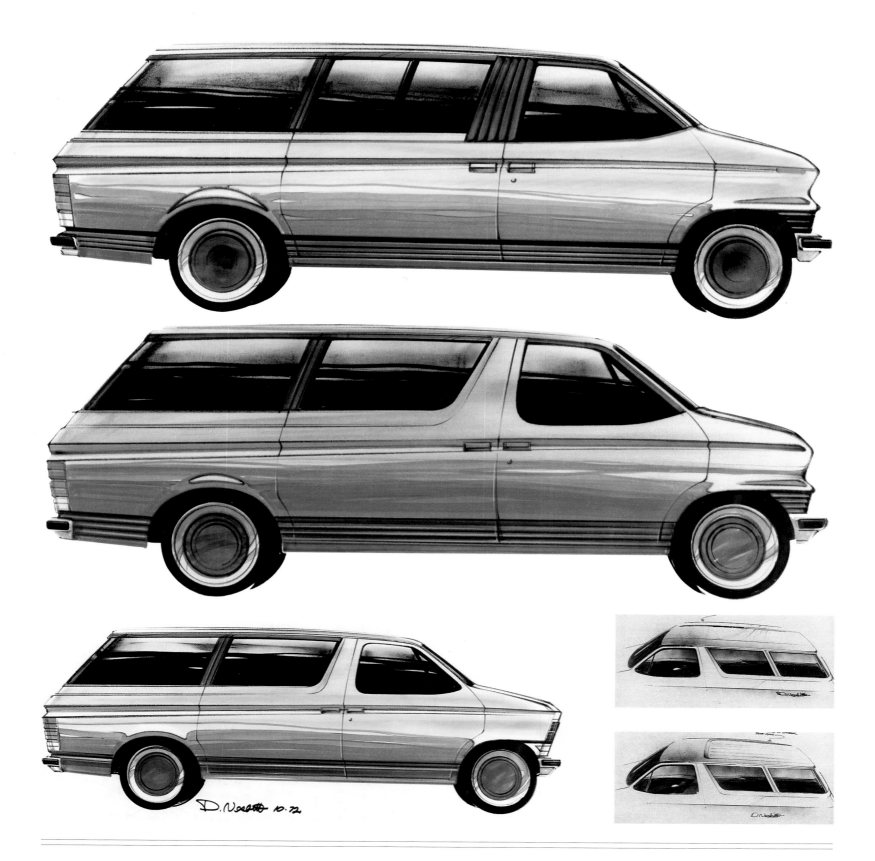

Although many proposals for the Carousel van were drawn up, all of the ones from this batch dated October 1972 kept the passenger-side sliding side door and the attractive slanted B-pillars. One version gave them a vented look (top), and did all of the remaining window pillars in black (except up front). The bodyside character line is pretty much horizontal in one rendering (top), but deliberately takes a gradual downward slant on two others (center and above left). This was again a device to make the transformed van look more car-like in its styling. The taillight treatments on all three proposals are virtually identical. Under the direction of designer Dick Nesbitt, a number of roof designs were explored for the Carousel (above right). Grooves and depressions in the roof panel not only enhanced appearance, but they made the panel stiffer and helped eliminate drumming to make the interior as quiet as in a luxury car. One roof proposal shown here simply has a depressed center rib, but the other has a recess in the rear two-thirds of the roof (with "hidden" ribs), which would have slightly lowered the position of an optional roof rack. This would have been another subtle design trick to make the Carousel more user friendly.

Gaylord: The Dilemma of a Dream

The end of World War II promised prosperity unknown to Americans since before the Depression. That promise was realized to a degree sufficient to encourage a good many dreamers and millionaires—not always one and the same—to try their hands at the car business. Most came up with pretty absurd results, confirming the view of *Road & Track* magazine founder John R. Bond that a little knowledge about cars can be right dangerous.

Brothers Jim and Ed Gaylord were different—rich to be sure, but devoted enthusiasts and serious about cars. Their father, who had invented the bobby pin and been smart enough to patent it, thus assuring the family fortune, had owned Marmons, Lincolns, Packards, and Pierce-Arrows. His sons grew up hot-rodding Packards, Cadillacs, and LaSalles. Speed engineer Andy Granatelli once built a Packard for Ed Gaylord that was the fastest thing on wheels in late-Forties Chicago—the police had the records to prove it. The brothers were also personal friends of Ed Cole, the master General Motors engineer, and spent many hours examining that company's latest experiments.

Jim Gaylord was the more visionary of the brothers, and in 1949 he met with Alex Tremulis, then finishing up as styling chief for the troubled, short-lived Tucker Corporation in Chicago. Tremulis recalled that Gaylord had just stopped by to talk cars, but the conversation ran long into the night. Five years later, Tremulis was working at Ford when Gaylord dropped by again. "Alex, I'm going to build the world's finest sports car," he declared, "and you're going to style it for me."

Tremulis refused, knowing his employers took a dim view of freelancing, but recommended Milwaukee designer Brooks Stevens, who would later build the Excalibur J sports cars and had ample experience with Alfa Romeo, Kaiser-Frazer, Willys-Overland, and American Motors. Jim promptly called Stevens, who immediately agreed to design the sports car.

What Gaylord wanted was a modern two-seat envelope body with classic overtones—namely an upright radiator and big headlights—plus a 100-inch wheelbase and even a retractable hardtop. Via several prior projects, Stevens had developed a respect for the Spohn works in Ravensburg, West Germany, and suggested that the prototype be built by that firm. Introduction was slated for the 1955 Paris Auto Salon.

Stevens's prototype was hastily completed and arrived just in time. A garish, finned affair with *de rigeur* wrapped windshield, it carried a vertical grille and big headlights, but in a front that was thrust rakishly forward at the top. Front fenders were angled to match, and were also cut back around the front wheels to leave them fully exposed. A contrast-color panel swept back from the fender hollows to the bottom and trailing edges of the doors, an effect Stevens called the "Washington Coach Door Line." This was finished in ivory to contrast with the black used elsewhere—Stevens' favorite color scheme.

A second, somewhat cleaner-looking car was soon constructed with smaller headlights but conventional fenders instead of the cutaway type. A retractable hardtop was again on hand per Jim's instruction, with a recessed rear window and integral air-extractor vents, thus pioneering "flow-through" ventilation. At the touch of a button, the rear deck lifted and a chain drive pulled the top back into the trunk. In this, the Gaylord was a full year ahead of Ford's 1957-59 Skyliner, which was considerably more complicated.

The chassis was Jim Gaylord's own design, a strong chrome-moly tube type with coil springs and A-arms in front and a beam axle on leaf springs at the rear, all very well damped. The first prototype carried a 331 Chrysler Hemi, but Cadillac V-8s were installed in two later cars (which also had full front fenders) as well as in one unibodied chassis. The chosen transmission was GM Hydra-Matic, modified so that full-throttle automatic upshifts came at peak revs.

The Gaylord's price was announced at a lofty $14,500, and was soon pushed to $17,500. Among the car's more unusual features were a spare tire that slid out from beneath the trunk on a tray, plus a vast array of instruments, each with its own warning light, set into a dash trimmed with oriental wood. Though it weighed nearly two tons, the Gaylord performed like a sports machine half as heavy. Top speed was 120 mph, 60 came up from a standstill in less than 10 seconds, and the fine chassis delivered smooth, stable handling on all kinds of roads.

But like others of its ilk, the Gaylord dream was not to be. Jim Gaylord was a perfectionist, satisfied with nothing less than exactly what he planned, and quality on the first two "pilot" cars didn't suit him. This led to a dispute with Luftschiffbau Zeppelin of Freidreichshaven, which was contracted to build production models in 1956. By early '57, the project was officially dead. Only four chassis were ultimately completed, one of which sported several beautifully color-coded components.

The main fault with the Gaylord enterprise was lack of solid production planning, something both brothers later admitted. But then, it's always easier to criticize in retrospect. What no one can deny is their brilliant chassis—virtually unbreakable and worthy of any sports car ever made. Styling suffered from period fads, but the basic shape was original and exciting. Remove the fins, the wrapped windshield, and the excess tinsel and you're left with a car that still looks good.

In their quest to build the world's finest sports car, brothers Jim and Ed Gaylord hired famed Milwaukee industrial designer Brooks Stevens to style a two-seater for them on a relatively short 100-inch wheelbase. The resulting car was strikingly modern on the one hand, yet carried overtones of classic-era cars on the other. The entire front end (top) was thrust rakishly forward at the top for an illusion of speed. It featured a narrow vertical grille opening reminiscent of the Classics of the Thirties. The grille was flanked by two huge headlights of the sort

one would expect to see on a prewar Rolls-Royce. The rear end (above left) was topped by moderately tall fins, which were coming into fashion when the car was designed. Taillights were nestled into chrome bezels that capped the ends of the fins. Vertically ridged bright trim suggested rear grillework, which housed additional rear lights. Dual exhausts exited just above the narrow rear bumper. Seven chrome strips rode on top of the rear deck, a feature DeSoto would adopt later for its hot Adventurer. Extremely deep side sculpturing (above right)

allowed for fully exposed front wheels and tires. This was made even more striking with additional bodyside sculpturing that created a sort of "Deusenberg-sweep" two-tone paint scheme, which Stevens preferred to call the "Washington Coach Door Line." Large wheelwells left the back wheels exposed as well. Knock-off hubs with twin "G" lettering highlighted the wheels, which sported a "vane" design. This first Gaylord, which was introduced in 1955, differed somewhat style-wise from what was to follow.

The first Gaylord had made its mark, but styling changes seen on the second car made the newcomer look a bit more conventional. This one (top), licensed as a 1957 model, lights the way with then-fashionable quad headlights, housed in this case within oval bezels. The eggcrate grille looked less fussy than the punched-full-of-round-holes texture used on the first Gaylord, and the chrome grille surround was bolder. A dual-stack front bumper allowed the parking lights to be moved from the fender tops downward between the bumper bars, and because the bumper was split, onlookers could see the entire grille opening. From the rear (above left), the revised Gaylord looked familiar, although the rear bumper had disappeared. In its place was a two-piece bumper that housed the exhaust outlets and left the entire center section of the car vulnerable. The biggest design change brought about more conventional front fenders (top and above right). No longer did they wrap far inward to completely expose the wheels; instead, the fenders flowed more smoothly into the other bodyside panels. Interestingly enough, the front wheel cutouts were almost square, seemingly at odds with the round opening that nearly radiused the shape of the rear tires. This was done to accommodate the Washington Coach Door Line two-toning, which itself had to be modified up front to be compatible with the new front fenders. But despite all the changes, there was no mistaking that the car was still very much a Gaylord.

The revised Gaylord kept its fins (top), hardly surprising in the 1956-56 era, not to mention its wraparound windshield, another styling device very much in vogue when the car was designed. One Gaylord feature, however, which had actually been seen on a Citroën way back in the late Thirties, was a retractable hardtop roof (center left). Citroën had been unable to capitalize on it, but Gaylord perhaps figured it could. And well it should have, for it could boast about a recessed rear window and

integral air-extractor vent, a pioneering version of "flow-through" ventilation. At a touch of a button, the rear deck lifted and a chain drive pulled the top down into the trunk. In this, the Gaylord was a full year ahead of Ford's 1957-59 Skyliner retractable, which was considerably more complicated. Inside (center right), the Gaylord went top drawer in the finest British tradition with a wood-faced instrument panel and fine leathers. The chassis (above) was of Jim Gaylord's own design, a strong chrome-

moly tube type with coil springs and A-arms in front and a beam axle on leaf springs at the rear. The first prototype carried a Chrysler 331-cid Hemi V-8, but two later cars, such as this one, ran with a Cadillac V-8. A modified Hydra-Matic made full-throttle upshifts at peak revs. Unfortunately, a $17,500 list price—$14,035 more than a base Corvette!—virtually guaranteed that the Gaylord would fail to attract enough buyers to make the car a commercial success.

Shocking Developments: GM's Experimental Electrics

Top: *GM's Research Staff devised these tiny two-seat experimentals for the company's "Transpo 72" traveling exhibit. Three "512" designs comprised (from far left) a gas/electric hybrid, a pure electric, and a gas-engine commuter; at the far right is the "511" three-wheeler,* *another commuter concept. Above left: A closer look at the 512 electric, which used an 84-volt battery pack; a separate 12-volt cell ran wipers, horn, and other accessories. The car was driveable with its cockpit canopy removed. Above right: GM's design staff devised this* *little number, called XP-833, again for Transpo 72. A small rear seat could allegedly hold two kids, while the underhood space was designed to accommodate a small gas engine, an electric motor, or hybrid gas/electric powerplant.*

Top left: *The '79 energy crisis prompted GM to test several Chevy Chevettes converted to electric power. Special lightweight materials and low-friction tires were used for maximum performance and operating efficiency.* Top right: *Though strictly a design study, this smooth-looking 1983 electric-car mockup envisioned near-term production using nickel-zinc batteries.* Above: *Unveiled in early 1990, GM's all-electric "Impact" two-seater is slated to see limited production in the mid- to late-Nineties; it should be virtually unchanged from the prototype shown here, which is said to do 0-60 mph in about eight seconds. A small-scale consumer evaluation program was underway in early 1994 as a first step toward retail sales.*

1979 I-H Scout Prototype: Trail to Nowhere

Targeted as a 1982 model, the '79 Scout prototype was evolved from (and maybe also built from) a similar "concept vehicle" shown in 1978. Differences involved the grille (like that of the jazzy post-'77 Scout SS), flared wheel arches, and solid instead of fabric doors. Chassis was standard 100-inch-wheelbase Scout, but body was made of various plastic-like composites. Production was precluded by a general decline in I-H sales, plus a labor strike and the sudden '79 recession triggered by a second energy crisis, all of which prompted I-H to abandon light-duty trucks entirely by October 1980.

Top left: "High-rider" styling made the '79 Scout prototype look meaner than the production SS or any previous Scout, but also vaguely like the Jeep CJ. Domed hood and multi-slot grille dominated the front. Middle right: All-vinyl bucket-seat interior was expected in upscale sport-utilities by the late Seventies and would likely have been standard for the planned new production '82 Scout. Top right: Hood cut spilled over to the front fenders for easier access to a 345-cubic-inch I-H V-8. Above: Rakish fastback roof with frameless glass hatch would have set the new Scout apart from the sport-utility herd. Doors would likely have been removable, as suggested by the simple exterior hinges. The prototype survives today at the Auburn-Cord-Duesenberg Museum in Auburn, Indiana.

Kaiser, Frazers, and Henry Js: New Ideas That Didn't Make It

Kaiser-Frazer Corporation lost about $100 each on some one million cars built over 10 years. That may sound like General Motors in the 1990s, but K-F lived in the boom economy of the late Forties and early Fifties, when poor judgment was about the only obstacle to automotive success. Which was precisely what forced K-F to abandon the U.S. market after attempting just three basic car designs. (K-F also oversaw the final cars of Willys, which it acquired in 1953, but that's a story for a later chapter.)

K-F was formed in 1945 as an alliance between Henry J. Kaiser, the rapid-fire construction and shipbuilding tycoon, and auto-industry sales veteran Joseph W. Frazer, who'd recently become president of moribund Graham-Paige. Both wanted to build new postwar cars. Frazer lacked the money but had ample experience; Kaiser had tons of money but no know-how. After discusssions, Joe Frazer signed on as K-F president, with Henry K. as chairman. After raising $12 million (total initial capitalization was $52 million), they purchased Ford's 2.7-million-square-foot wartime bomber plant at Willow Run, Michigan—the world's largest factory under one roof—and hired design and engineering talent by the boatload. As production got going in June 1946, K-F people felt, as one official later recalled, "as if there wasn't anything we couldn't do."

And indeed, K-F could do no wrong—for a while. Though conventionally engineered, the debut 1947-48 Kaisers and Frazers were fresh and appealing against most rivals' warmed-over pre-war cars. Prices were stiff, yet K-F vaulted to eighth in industry sales within a year, prompting some to call it the "postwar wonder company." Yet though all seemed amicable between Joe and Henry, Kaiser soon started replacing Frazer's people with his own. By 1949 he controlled everything from the boardroom to the factory floor.

Which is how Henry was able to mandate 200,000 cars for 1949 over Joe's vehement objection. Knowing that K-F could offer only facelifts against mostly all-new Big Three designs, Frazer wanted to build fewer '49s, then come back strong with stunning all-new models then being planned for 1950. But at one memorable board meeting, Henry declared, "The Kaisers never retrench!" With that, Frazer resigned and Henry appointed his own son Edgar as president. Henry got his 200,000 cars, but sold only 60,000 and lost $30.3 million, then close to a U.S. record for a company of that size.

This miscalculation had two ultimate results: the end of the Frazer nameplate and a six-month delay in launching the new second-generation Kaiser, which bowed in spring 1950 for model-year '51. Until then, K-F pushed out '49 leftovers very slowly despite big discounts. Some were reserialed as "1950" models; about 10,000 others were transformed into '51 Frazers via new front and rear sheetmetal.

It was about all K-F could do, but the company was never short of ideas on how to make the old stuff seem new. These came mainly from K-F Styling principals Bob Cadwallader, Herb Weissinger, and Arnott B. "Buzz" Grisinger, all recruited from Chrysler, plus the able Cliff Voss and Milwaukee-based consultant Brooks Stevens. Between them, these five conjured countless facelifts on the original '47 bodies, plus variations including hardtop coupes, two-door convertibles, fastbacks, and even wood-trimmed sedans. Two pioneering styles, the hatchback sedan and four-door "hardtop," did see production in the 1949-50 Kaiser line (as the Traveler/Vagabond utilities and hardtop Virginian) and as '51 Frazer models. Kaiser and Frazer also offered America's first postwar four-door convertibles, but these were just cut-down sedans done on the cheap.

Like K-F's first-generation cars, the look of the rakish new '51 Kaiser was largely owed to the renowned Howard A. "Dutch" Darrin, another consultant who carried the day over proposals from both Stevens and the in-house team. Weissinger and Grisinger finalized things like bumpers, hood ornament, and grille, but the long, low shape was pure Darrin. Sleek and beautiful, the 51 Kaiser had no design peer among Detroit sedans for a good five years.

Because Joe Frazer and his namesake car were still around, the new Kaiser was also planned as a Frazer—again, a more luxurious and expensive version with slightly different styling. Weissinger, who also supervised the '51 Frazer restyle, envisioned a complex eggcrate grille a la 1947-50, placed low on the new Darrin body. Toward 1949, however, it was decided to postpone the second-wave Frazer until 1952. At one point, Weissinger tried grafting the '51 Frazer front onto the new Kaiser shell, which would have been ghastly. But none of this mattered in the end. With Joe Frazer about to leave after being reduced to the meaningless position of board vice-chairman, the Frazer line was deemed unnecessary after '51 and did not return.

Meantime, the so-called "Anatomic" Kaiser managed a strong 139,000 sales for 1951, (helped by a six-month jump on the competition), only to score just 32,000 for '52 despite a heavy facelift by K-F Styling (with many more Stevens ideas left unused). Sales skidded again for '53, thudding at 28,000. There were two big problems. First, Kaiser still had only an anemic six to counter the potent overhead-valve V-8s of most rivals. K-F was working on a V-8, but couldn't afford to produce it because of the second problem: money squandered on the unhappy Henry J compact, of which more later.

To some extent, the "Anatomic" also suffered from offering just two- and four-door sedans, hatch and non-hatch. Not that K-F didn't consider other body styles while finances were healthy; a proposed hardtop coupe, dubbed "Sun Goddess" by stylist Alex

Back in 1947, when Kaiser-Frazer was enjoying great success in the booming postwar auto market, Milwaukee-based industrial designer Brooks Stevens pitched a small car to K-F. Designed around a 108-inch wheelbase (same as the 1956 Rambler),these two renderings (top row) give a good idea of how the car might have looked. Despite slab-sided bodies, styling was actually a bit dated and somewhat British-looking. This car nevertheless would probably have been a saleable proposition. Visibility, for example, was excellent for the time, and price and economy would have been strong selling points. The ever-optimistic Stevens also moved forward with suggestions for updating the original 1947 Frazer. One of them, an upmarket "Town Sedan," pretty much left the sheetmetal unchanged (center), but sported lower-body cladding that predicted the use of anodized aluminum trim in the second half of the Fifties.

Similar lower-body two-toning would also be trendy in the Eighties. An apparently cheaper model, the "Custom Sedan" (above), forsook the chrome-capped parking lights and had less bright trim above the grille bar, but looked a bit taxi-like in three-tone paint. Note the mildly heart-shaped windshield on both cars. This would become a design hallmark (and a safety feature) of the completely restyled '51 Kaisers.

Tremulis, was actually constructed from a '51 two-door. Basically stock from the beltline down, it carried an attractive pillarless roofline with broadly wrapped backlight. A convertible was also discussed, and would have looked great. But again, there just wasn't enough money.

Kaiser was thus forced to rely on facelifts, plus interesting trim options like the colorful "Dragon" series, to get through each year. Somehow, though, money was found for a 1954 update boasting a wide concave grille, wrapped rear window, and three-sided "Safety-Glo" taillights with supplemental red lenses atop the fenders. Herb Weissinger was again responsible for a remarkably adept K-F restyle.

But by that point, time had run out. Despite a new performance-boosting supercharger option, Kaiser sold just 8539 of its '54 models. After struggling to build just 1291 of the virtually unchanged '55s, management decided to abandon the U.S. passenger-car market to concentrate solely on Jeep vehicles (acquired with the Willys takeover). The final proposed Kaiser facelift was a garish, two-toned affair for 1955. Also left stillborn was a complete makeover planned for '56.

Yet the Kaiser nameplate was far from finished. Soon after quitting the States, the indomitable Henry Kaiser visited Argentina to talk with dictator Juan Peron about starting a local auto company. This became Industrias Kaiser Argentina AS (IKA), which was put under James McCloud, Edgar Kaiser's brother-in-law. From 1958 through 1962, IKA sold a 1954-55 Kaiser Manhattan as the Kaiser "Carabela" (for caravelle, the ship) at the rate of about 3000 a year. Save minor trim changes and a suspension toughened to handle rough Argentine roads, it was identical to the last American Kaisers right down to its 115-horse-power 226-cubic-inch flathead six. The supercharger option wasn't offered. Neither was automatic, the only transmission available being a three-speed manual.

It's a tribute to Darrin's "Anatomic" styling that the Carabela lasted so long. It might have lasted even longer, for ideas were afoot as late as 1960 to give it new life. That's when no less than Darrin himself was asked to devise a facelift. He produced two concepts, one mild, the other wilder. The more conventional involved just a modestly lipped windshield header, ponderous front fender/door moldings, and a chrome strip run back from the front wheels above the rocker panels. Darrin mocked this up on an early '54 Kaiser Special (which lacked the wraparound rear window of the "late" '54 U.S. models). The more ambitious proposal would have looked very nice indeed. This involved new front sheetmetal with lower fenders and hood sloping down to a

broad U-shaped grille with a simple horizontal bar, flanked by quad headlights. Management was favorably disposed, but decided sales were insufficient to warrant the tooling expense.

Still, the Kaiser wasn't dead. Back in Toledo, where Henry had repaired to build Jeeps after selling off Willow Run, James Anger of Product Development had concluded that only the Carabela's superstructure needed updating. Envisioning a squared-up "formal" style like that of contemporary Ford Thunderbirds, he actually constructed a prototype using an old Manhattan sedan, modeling the new roof in fiberglass and side windows in Plexiglas. If something of a mismatch against the rounded lower body, the new top didn't look too bad and achieved a considerable increase in glass area, which was already good.

But as in America, all these "extensions" were doomed for lack of sufficient sales volume to justify tooling costs, and the Carabela was dropped after 1962 because the old dies had simply worn out. Just before Henry Kaiser sold his interests to the locals in 1965, IKA began selling a facelifted '64 Rambler American, called Torino, which enjoyed good success into the Eighties. IKA later built Renaults under license and was eventually acquired by that French automaker. Renault later sold out to Ford Argentina, which thus inherited the locally built civilian versions of the military Jeep, similar to what Ford Dearborn had built during World War II—proving, perhaps, that what goes around, comes around.

Now, let's go back to early 1950, when K-F introduced the Henry J as America's second postwar compact. Though far less successful than Nash's Rambler, which arrived a bit earlier, the Henry J at least stabilized K-F's then shaky finances for a time. Henry Kaiser had promised a new car all Americans could afford, which is why he was able to borrow $69 million from the Reconstruction Finance Corporation in 1949. While $25 million was destined to finance K-F's heavy inventory of leftover '49 models, Washington okayed the loan because some $12 million was earmarked for the new compact.

The Henry J promised sturdiness and low operating costs, which it delivered. But like other K-F models, it was, for various reasons, relatively expensive—only a little less costly than a "full-size" Ford or Chevy. That made sales tough, and the market was quickly satisfied anyway. From a healthy 80,000 for '51, volume plunged to just 1123 by 1954, after which the model was dropped.

Styling was also a factor, for the Henry J was anything but lovely: a pudgy-looking two-door fastback sedan with little Cadillac-style tailfins and a front-end vaguely like the '51 Frazer's. Yet it might have been much prettier. Dutch Darrin had proposed something like his sensational '51 Kaiser, which was being evolved at the same time, with similar "Anatomic" styling on the 100-inch wheelbase. Darrin built a prototype at his Santa Monica, California, studios by sectioning 18 inches from a '51 Kaiser club coupe. Though it looked far better than what appeared in show-rooms, management felt the Henry J should look "new"—meaning different—and thus chose the ungainly 1951-54 styling, which actually came from a K-F supplier.

Briefly, K-F had great plans for the Henry J, including a convertible and hardtop coupe based on the lone two-door style. Many proposals were advanced, but none reached the assembly line, though a few dealers built convertibles out of sedans. Also considered—and quickly abandoned—were a two-door station wagon (which would have looked quite neat) and a four-door sedan (whose rear doors would have squeezed everyone except toddlers).

Had it survived, a redesigned Henry J would have appeared for 1955 per plans made in 1950. According to former K-F managers and company documents, it was intended to last until 1959 or '60. The most radical of the proposed designs was the "105," conceived for that wheelbase length by the free-thinking Alex Tremulis, recently involved with Chrysler and the Tucker. Tremulis was a champion of streamlining, and believed it would make the new Henry J truly revolutionary. "Our proposal was between the [original and the 1954] Kaiser-Darrin [sports car] in size," he later said, "but with its lightness and small frontal area, it could outperform both. We figured it to return 25 mpg and yield

Opposite page: *A clock-face steering wheel designed by Brooks Stevens for a proposed Frazer. This page: Three proposed instrument-panel designs (top row) indicate that Kaiser-Frazer was serious about safety. Each featured a padded top and a lower edge that sloped away from the driver and passenger to help protect knees in an accident. The speedometer and gauges were placed directly behind the steering wheel for quick, easy reading. The radio and* speaker resided in the center of the panel for good sound distribution. Brooks Stevens sketched proposals for updating the original Kaiser and Frazer front ends (second row). As he saw it, the Frazer would have a fairly elaborate bumper, and just above it a wide horizontal slot in place of a grille. The Kaiser, depicted as a Custom, would follow the same theme, except that the slot would be incorporated into the bumper itself. Some ideas from Stevens for 1950 (third row): more* wide-slot grille openings, a lower (but still rounded) hood bulge, and vastly increased window area. Note in particular the wraparound rear window. Still more ideas for 1950 (bottom row): a Kaiser Custom four-door with a recessed vertical-bar grille and a huge central bumper guard; a two-door with a "loop" bumper/grille of the sort popularized by Chrysler Corporation in the late Sixties; a four-door station wagon.*

an estimated top speed of over 100 mph. The weight was only 2500 pounds." The basic coupe had broad areas of glass, sharply undercut front fenders, and a modest grille, plus far more leg and head room than the original Henry J. Tremulis later called the 105 "another Tucker—years ahead in concept and function. If it had been produced, in my opinion, there would have been a Big Four."

Darrin, meantime, never stopped pushing his own Henry J ideas, and some were actually mocked up. He also won a victory of sorts by convincing Kaiser to build the fiberglass-bodied two-seat roadster he'd designed for the Henry J chassis. This bowed for 1954 as the Willys-powered Kaiser-Darrin ("KDF-161"), with novel sliding doors and a three-position soft top, two patented Dutch innovations. Only 435 were built before Kaiser fled the U.S.

In the end, K-F was doomed by its '49 folly. Though its later cars were good and sometimes innovative, none were able to impress the public enough. But as Edgar Kaiser once said, "Slap a Buick nameplate on it and it would sell like hotcakes." He was probably right.

The ever-creative Brooks Stevens created numerous styling proposals for the 1950 Kaiser-Frazer product line. Like the cars seen on the previous page, all those seen here sported thin-pillar styling and lower bodyside trim. A two-door (top), looking much like a convertible, neatly merged the lower body trim into the rear bumper, and topped that with tall vertical taillights. Fastbacks were still popular when Stevens drew his styling studies, as shown by the pair of two-door models seen here (second row). The one at the right hints at the roofline of the all-new '51 Kaiser. A more upright and formal-looking two-door sedan (above right) was really more in tune with the times by 1950; even Chevy would drop its fastback after 1952. Although Kaiser-Frazer never produced a station wagon (above left), that body style was explored many times by various designers, both within and outside Kaiser-Frazer. This Brooks Stevens proposal mated a woody look with all-steel construction. Stevens had of course already done this with his 1946 Willys Station Wagon, which beat the all-steel 1949 Plymouth Suburban two-door wagon to the marketplace by three years. It's unfortunate that K-F never managed to produce a wagon—or a hardtop, for that matter.

Top row and center left: *Snapped in the Kaiser-Willys styling department at Toledo circa 1960, these rare photos record the facelift proposed for the Argentine Kaiser Carabela by designer Dutch Darrin. Key elements included a modernized face with quad headlamps, flatter hood and rear deck, different taillamps, and elimination of the* "sweetheart dip" *from windshield and backlight.* Center right: *The same full-size model was later reworked to test a slightly different trim treatment. Note the absent rear fender skirts.* Above: *K-F considered non-sedan body styles for the second-series Kaiser. "Sun Goddess" hardtop (left) was proposed by staff designer Alex Tremulis, while the great* Thirties coachbuilder Ray Dietrich *oversaw the conversion of several '51 club coupes (some sources say as many as six) into prototype convertibles (right). Unfortunately for K-F, it never had the money to offer these, nor a V-8 that was much-needed for competitive performance, though the firm had one such engine in the works.*

Lost LaSalles: Keeping the Spirit Alive

Cadillac's LaSalle was one of only two late-Twenties "expansion makes" to survive the Depression. The other, of course, was Pontiac. But LaSalle did not survive the Forties, despite moving downmarket from "near-luxury" junior Cadillac after 1933 to an upper-medium-price product that enjoyed several good sales years.

The reason for LaSalle's eventual demise was as close as any Cadillac showroom. By 1940, LaSalle was nearly identical in form, finish, body styles, and performance with Cadillac's low-line Series 62, and both were priced close to senior Buicks, which had its eye on the same market territory. For General Motors, it made no sense to continue LaSalle as an in-house rival to those better-established brands.

A more critical factor was the slow but steady upturn in demand for high-luxury cars that accompanied the halting general economic recovery in the years leading up to World War II. Taking due note, Cadillac made a strategic decision that would take it to the very top of its class: namely, to abandon the upper-medium field and return to nothing but all-out luxury. This contrasted with Packard, whose continued over-reliance on cheaper models after the war cost the make its blue-chip image—and eventually its life. As one executive said later, Packard just "handed the luxury market to Cadillac on a silver platter." So in the end, LaSalle fell victim to changing times. Today we'd say it was a niche model that lost its niche.

Yet before its demise, a 1941 LaSalle model was in the works (three-year lead times then being the Detroit norm, remember). This was a heavy facelift of the lovely 1940 design, itself virtually all-new. Headlamps again nestled firmly within the fenders, and the slim horizontal-bar LaSalle radiator returned per recent tradition. But where the 1940 had curved front-fender "catwalk" aprons, the proposed '41 was more blunted. Prominent vertical slots again flanked the radiator, but were shorter in keeping with the industry trend to more horizontal "faces." The profile was pure period GM. Front-fender trailing edges were newly squared off in the manner of Cadillac's stylish Sixty Special. Rear fenders wore fashionable skirts with a circular emblem where the hubcap would have shown had the wheels been exposed; some production '41 Caddys had this too. Chrome trim included triple strips at the base of all fenders and sparkling appliques along the vestigial running boards. GM Design chief Harley Earl believed this made for a brighter look on used-car lots that helped prop up resale values, which is why so many '41 GM cars had a "shiny" look.

The '41 LaSalle went as far as a pair of full-size Series 52 mockups, fastback and notchback, both four-doors. With the decision to axe the line, Cadillac substituted a new Series 61 of comparable size and price. A decade later, it was gone, too, as the division moved even more decisively toward luxury-class leadership.

Despite its role as Cadillac's less-expensive companion, LaSalle was a great loss for many GM aficionados, who still thought of it fondly as years passed. After all, designing the first 1927 LaSalle was what had brought Harley Earl to GM, where he set up the famed Art & Colour Studio as the industry's first in-house styling department. The '34 LaSalles were far less prestigious than earlier models, but were remembered more for somewhat daring looks than middling price.

With all this, it's not surprising that some GM designers and even executives continued to have visions of LaSalle's eventual return as a specialized Cadillac. Prime among them was Harley Earl, who put the name "LaSalle II" on two exercises for the 1955 edition of GM's traveling Motorama show. Both wore vertical-slot grilles echoing the aborted '41, flanked by vertical bumpers bearing big "bullet" guards and wrapped around to the sides. There were also "LaS" emblems as used in LaSalle's last years.

Still, these concepts were dissimilar to each other. One was a flashy two-seat roadster of the Corvette stripe, with elliptical bodyside concavities like those destined for the production '56 'Vette. Stubby rear fenders were abruptly cut off to leave the wheels exposed (something like the Brooks Stevens treatment for the front of the stillborn '56 Gaylord). Chassis side rails housed the exhaust pipes, which exited just ahead of the back wheels.

The other '55 LaSalle II was a hardtop sedan with rear-hinged back doors, one of the few throwback touches Earl indulged in. (The production '57 Cadillac Eldorado Brougham would have them too.) Seating for six was provided despite a compact 108-inch wheelbase. Overall length was just 180 inches, height a mere 50 inches. That lowness was partly achieved with 13-inch tires, rare for even period Detroit showmobiles. Predictive features included unit construction, a big compound-curve windshield similar to 1959 production design, and an experimental small-block aluminum V-6 that GM was toying with at the time. Concave bodyside ellipses, again finished in a darker hue, were shared with the roadster (as was V-6 power), but rear wheels were only semi-exposed in "jet tube" fenders a la the '53 Corvette.

GM publicity described the LaSalle II hardtop as "a new concept of passenger sedan styling directed to recapture the distinctive exclusiveness and high quality of craftsmanship of the original LaSalle." But to many, it just looked silly. It also looked much like other recent Earl "dream cars" including several Cadillac concepts, the '55 Chevy Biscayne, and the '56 Impala Sport Coupe. The LaSalle II roadster wasn't really fresh either, having been foreshadowed by the 1954 Buick Wildcat II, Olds F-88, and Cadillac La Espada/El Camino—not to mention the '53 Corvette. But then, both LaSalle IIs were strictly for show and never intended for showrooms.

Photographed in late January 1940, this notchback four-door (top and above) was one of two full-size models built for evaluation of proposed 1941 LaSalle styling by General Motors executives. Resemblance to other period GM cars is *obvious in the mid-body region and at the rear; strong vertical-theme front treatment, however, would have been unique in GM's '41 fleet. Years later, GM design chief Harley Earl spearheaded* *the LaSalle II show cars (center row), seen at the 1955 Motorama show. The roadster evoked Corvette, while the hardtop sedan epitomized sporting elegance.*

Production was definitely on GM's mind when the LaSalle name resurfaced a few years later in connection with the project that produced the 1963 Buick Riviera. At first, this Thunderbird-beater was proposed as a new personal-luxury Cadillac line called LaSalle, and several body types were developed, including convertible and hardtop sedans. The four-door droptop would have been quite timely against Lincoln's then-new Continental model, but Buick's poor sales in that period dictated some added product help, so the car was assigned to Flint and offered only as a hardtop coupe. Thus ended the first chance for a new LaSalle since 1940.

But the name continued to exert considerable magic within the GM halls of power, and another chance came in the early Seventies. Cadillac was planning a new small sedan, and there were serious thoughts of calling it LaSalle, though "Leland" was also in the running (honoring Cadillac's founder, Henry Martyn Leland). LaSalle, however, was all but assured—until a division executive came across an article that characterized the original line as "Cadillac's only failure." That was enough for the sales force, which voted for a newer, more recognized name with some success behind it: Seville. Reportedly, LaSalle was also contemplated for what became the subcompact Cimarron of 1982-88, but given that car's unhappy record and "loser" image, this third-time rejection was a definite charm.

People still talk about LaSalles, but they're now mostly historians and old-Cadillac fanciers. Will the name ever be revived? Probably not, though you can never can tell. Let's just hope that if Cadillac ever does see fit to bring it back, it will be for a car truly worthy of the romantic originals, lest the name be forever sullied. Otherwise, LaSalle is probably best left in limbo. As they say, some things just can't be duplicated.

Cadillac designers ran up this four-door fastback for styling review of the '41 LaSalle. Like the notchback, it was done up in upper-level Series 52 trim, as indicated by the front license plate (above left). Squared-off front fenders and rear fender skirts with a tiny "hubcap" bauble (top) provided a family resemblance with senior '41 Caddys, as did the general fastback contour (above right). Erosion of a once-clear price niche and renewed demand for costlier, pure-luxury cars prompted Cadillac's decision to drop LaSalle after 1940.

Top and center row: *Likely considered for 1940, not '41, was an intriguing LaSalle version of Cadillac's handsome and predictive new 1938 Sixty Special, designed by a young William L. Mitchell. This mockup, photographed in mid-April 1937, wears a 1940-style LaSalle radiator, but also squarish headlamps rather than the round sealed-beams* actually used. Though LaSalle was selling well in '37, management likely vetoed this model for fear of sales interference with the "real" Sixty Special. Above left: The '41 LaSalle fastback mockup was remodeled at some point—perhaps by an artist's airbrush—to show planned trim for the lower-line Series 50 models. The *differences are hard to spot. Above right: The same mockup in another studio photo (courtesy of former GM designer David Holls), but refinished or retouched to test a monotone paint scheme and darker colors. The early-1940 dates on all these pictures indicate the decision to drop LaSalle came very late in GM's 1941 model planning.*

Nixed Nash: The Pininfarina Prototype for 1955-56

Italy's renowed Pininfarina coachworks had contributed its masterful design acumen to all manner of early-Fifties Nashes. Soon after the Nash/Hudson "Hash" merger in early 1954, the new American Motors Corporation commissioned PF for an all-new big-car design, eyeing a possible 1955 or '56 introduction. The result was the prototype four-door sedan seen here, with styling carefully evolved from the first "Farina Nash" design of 1952-54. Differences began with a wrapped windshield (then the height of fashion), smoother lower body contours, and a sleeker rear deck. Happily, this car survives today in the hands of a private collector.

Center: *From behind, PF's '55 Nash prototype looks somewhat like contemporary Studebaker sedans.* Top right: *A period PF photo shows the car with its original windshield divider bar that has since been removed.* Top left: *Small "trafficator" turn-signal beacons pop up from each C-pillar, an expected European touch.* Top center: *As on 1952-54 Nashes, PF insignia is prominent on the lower front fenders. "Speciale," of course, means prototype.* Above left: *Interior was spacious. "Step-down" floor took a cue from the last "pre-Hash" Hudson series. Dash looks similar to Nash's final '55 design.* Above, upper right: *Larger and more blunted fenders than Nash ever used flank a neat oval grille with inboard headlamps, a motif PF had introduced on the 1952-54 Nash-Healey sports car. Nash adopted a concave oval grille of similar size for 1955-56. Note the relatively small hood.* Above, lower right: *Crossed-flags rear-deck insignia signified another Nash/Farina collaboration, but only Kenosha's badge appeared up front.*

Unseen Styling from Nash, Hudson, and AMC

After three years of tubby Airflytes, Nash completely restyled for 1952 along lines suggested by Italy's Pininfarina design house. But though production models wore "PF" badges, the final design actually owed more to Nash's own Ed Anderson, who was responsible for these three workouts, among many. Skirted wheels, wrapped rear window, and dipped beltline were picked up from PF's proposal, but more aggressive, lower-set fronts were tried (top) along with variations in headlamps, rear end, and side trim (above left). Toward the end (above right), detailing and C-pillar treatment were both greatly (and pleasingly) cleaned up.

Top row: AMC did up this neat thin-pillar four-seat wagon around 1956 as a possible companion for the two-seat Metropolitan convertible and coupe. Back seat left no room for luggage, but presumably flopped down into a proper platform, which could also be reached through the flip-up back window.

Mediocre Met sales likely figured in the decision not to proceed with production. Other photos: A quintet of clay scale models from 1955 shows AMC's thinking for 1957 Nash and Hudson styling, which assumed a new shared platform with a 125-inch wheelbase. But that must have looked a "dinosaur" to

new AMC president George Romney, for it was vetoed in favor of the more compact Rambler-based Ambassador that effectively replaced both Nash and Hudson for 1958. It was just as well, for these mockups were full of period styling clichés that buyers would soon reject.

Top row: *A pair of elevation drawings from April 1955 anticipate the '57 Hudson. Long-tail styling (left) has a certain period Mercury flavor. Horrific "V-Line" 1956 styling would have been even worse (right) had some ideas gone through. Second row: A two-sided scale clay from August '55 shows tail and roof* workouts for '57. The right-side treatments were doubtless planned for Hudson, the left for Nash. Other photos: Undated, but probably from later in '56, this more fully finished scale model proposed wild triangulated shapes all over AMC's '57 senior models, plus hidden headlamps. Bold Ford-like side *trim was earmarked for Nash (above right), while the Hudson version was graced with a prominent triangle nose logo (main photo) and rear-fender hashmarks (above left). Vee'd C-pillars suggest actual 1957-58 Lincoln-Mercury styling.*

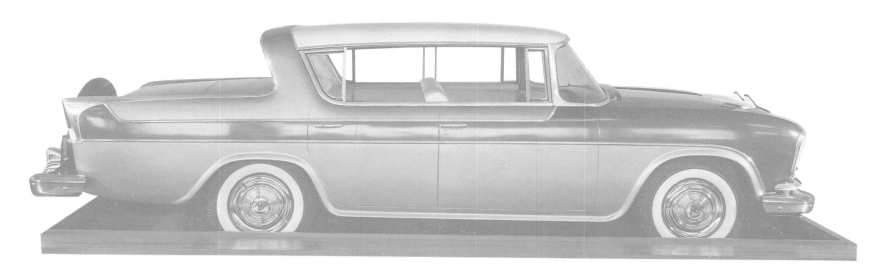

Top two rows: *This full-size mockup from 1955 proposes a rather busier '57 Rambler facelift than what finally appeared. Note the different fender treatments on the pillarless right side versus the pillared left, plus the reverse-slant backlight (possibly a drop-down affair) that foreshadows the '58 Continental Mark III but was not* adopted. Other photos: *Just before deciding to drop Nash and Hudson, AMC planned this styling for 1958 models built on an 11-inch stretch of the 108-inch-wheelbase Rambler platform, with all the extra inches ahead of the firewall. This is the Hudson version, which went from a simple bar grille (third row from top, left) to a fine-* checked affair. *Both this and the vee'd bumper guard were saved for the replacement "Ambassador by Rambler" that was offered instead. Intriguingly, these shots seem to date from mid-1957, suggesting the decision to axe the old nameplates was made very late in AMC's '58 model planning.*

Had the make continued, Hudson would have been more ambitiously redesigned for 1958, as shown in this proposal from earlier in '57, but it would still have been built as a stretched Rambler and not as a continuation of the 1955-57 Nash platform. Presumably, there would have been a Nash version as well with markedly different styling. Front end and side trim here continued recent Hudson "cues," while the canted gullwing-type fins marked a departure that suggested an exaggerated '57 "Packardbaker" rear end.

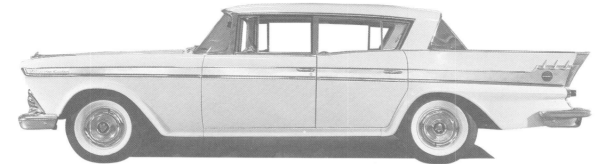

Top: *As noted, AMC hoped to improve its big-car fortunes after '57 by dropping Nash and Hudson in favor of Rambler-based Ambassadors offered in more body styles, but these models fared no better at first. The side-trim workout in this undated photo was likely for 1959, indicated by the '58-type greenhouse and the smoother beltline rise from front door to tailfin, both of which were featured in production.* Second row: *Though AMC didn't offer a mid-size convertible until*

1965, styling chief Dick Teague contemplated one for '64 along with his facelift of the previous year's new "Uniside" Classic/Ambassador designed by predecessor Ed Anderson. This model wears "Cavalier" hood script, a name that Teague revived for one of his 1966 "Project IV" show cars. Third row: *Looking rather Chrysler-like from astern, this facelift of the '64 Ambassador was proposed early on for 1965, but was ultimately discarded for a bigger, cleaner,*

more rectilinear look. Above: *Teague fought hard to keep the Javelin-based AMX despite unimpressive 1968-70 sales. This mockup, evidently shot in November 1969, tried mating the original AMX body with the humped-fender front "clip" from Teague's restyled '71 Javelin. In the end, AMC couldn't afford to continue the two-seater past 1970. Note the big hood scoop, tiltable rear spoiler, and side exhausts—all racy but fanciful fillips.*

Rocket Action: Oldsmobile's Fabulous Fifties Dreams

Though strictly for show, the four-seat Starfire convertible of 1953 provided an accurate sneak preview of near-term Olds styling. Wide "fish-mouth" grille would appear in modified form on Lansing's '56 models, while the wrapped "Panoramic" windshield and "hockey-stick" side trim were destined for showrooms in 1954. Tri-spinner wheel covers were shared with the limited-edition '53 Fiesta and would be widely copied as aftermarket accessory items. Like the '53 Corvette, Starfire's body was made of that wondrous new material called fiberglass, but an Olds "Rocket" V-8—tuned for a smashing 200 horsepower—sat beneath the hood.

Top and center: *Lansing's 1956 "car of tomorrow" was the Golden Rocket, a wild two-seat coupe striding a trim 105-inch wheelbase and measuring 201 inches long overall. Styling was very "Buck Rogers," from prow-shaped nose to stubby fins and bullet taillamps. "Flipper" roof panels raised electrically when the doors were* *opened to assist entry/exit in this low, gold-painted projectile, which became the styling takeoff point for the '58 Corvette. Split rear window would appear much later on the '63 Corvette Sting Ray coupe. Above: 1954's F-88 was a two-seat convertible like the original Corvette, right down to an identical 102-inch* *wheelbase. Styling, of course, was all Oldsmobile, and again heralded near-term showroom styling. Though many "dream cars" were cut up once their purpose was served, this one-of-a-kind survives today in like-new condition in the hands of a very lucky private collector.*

GM was in the habit of "facelifting" some dream-car designs just as it would for everyday production models. Here, for example, is a 1957 experimental that's also called F-88, built three years after its Motorama namesake. Except for a wrapped windshield, general Oldsmobile look, and two-seat roadster format, it looks completely different from the '54 iteration—also much less Corvette-like, though the protruding grille and quad headlamps (top) give faint hint of the Chevy sports car's face for '58. Profile (above) emphasizes a fairly restrained design for the period.

Another look at the second F-88 from 1957. Snazzy interior (above left) was very similar to that of the original '54 Motorama design. Note the gauges running vertically from the middle of the dash down into the console. GM styling chief Harley Earl was ever a fan of aircraft motifs, hence this contrived cockpit treatment and the modest, abbreviated fins striding the rear deck (top). "Jet-exhaust" taillamps were a natural tie-in to Lansing's "Rocket" V-8 and period advertising themes. Bold "Oldsmobile" grille nameplate (above right) is almost unreadable. Even less legible lettering would be used on the front and rear of Oldsmobile's production '59s.

143

Last Days in the Bunker: Packard's Plans for 1957-58

In its day, the main Packard plant in Detroit was among the most beautiful and well-ordered of factories, its spotless Packard Gray interior staffed by an experienced workforce of high morale. The facade, easily seen from tree-lined East Grand Boulevard, was studded with impressive pillared entrances, each bearing the Packard name in dignified block letters; the effect was not unlike that of a mausoleum.

A mausoleum it was in the summer of 1956, when production ground to a halt after years of customer flight to Cadillac, Lincoln, and Imperial. Hardly anything was left except the styling studio, still valiantly working on "real" Packards for 1957-58. The late Richard A. Teague, then a staff designer and ever a car enthusiast devoted to the marque, later called this period Packard's "last days in the bunker."

Financial resources dried up one after another that summer. Gradually, big East Grand was emptied when workers were shifted to Studebaker's South Bend, Indiana, factory, where Studebaker-Packard Corporation planned to build Studebaker-based Packards amid a frank cash crisis. In the authoritative *Packard: A History of the Motorcar and the Company,* Teague recalled that "Styling was the last to go because [management] thought there was [still] some chance. You knew goddamn well the end was close, but you kept hoping for the life raft. Rumors? You wouldn't believe the rumors. . . . Everybody from Universal CIT to Ford was buying us out."

Left behind in Detroit was the true '57 Packard, not a pretender in Studebaker dress but the car with which S-P president James Nance hoped to rebuild the luxury end of his business. Besides the usual numerous renderings and clay models, there were several full-size mockups and a running "mule," all inspired by the distinctive Predictor "dream car" that had toured the show circuit earlier in '56.

Built by Ghia of Turin, the Predictor was executed under Packard design chief Bill Schmidt, but strongly reflected Teague's thinking. Advanced features included a windshield that wrapped up as well as around, quad headlights hidden behind clamshell doors, fenders level with the hood and rear deck, and a square, chiseled shape. Also on hand were several ideas from recent Teague-styled Packard show cars: reverse-slant retractable backlight (previewing the '58 Continental Mark III), shapely "cathedral" taillights, and smart ribbed bodyside moldings that ran from the doors right around to the front. That trim ended abruptly to frame a slim vertical nose with Packard's traditional "oxyoke" radiator shape, which Nance had lately been trying to resurrect as a sales-booster. (It might have been used for '54 had time allowed, though Teague managed something far better for the one-off '55 Request). "Rolltop" roof panels slid away to ease entry/exit in what was a pretty low car; they could also be left open for ventilation—a kind of embryonic T-top.

Inside, the Predictor was all convenience, with electronic pushbutton Ultramatic transmission; electric servos for decklid, roof panels, and windows; and individual contoured seats with reversible cushions—leather on one side, fabric on the other, as on Packard's '56 Caribbeans. Roof sail panels wore portholes like those adorning contemporary Thunderbirds, plus courtesy lights and a jeweled escutcheon. The decklid was adorned with a large "circle-V" emblem created by Teague with hopes of establishing another "timeless" automotive symbol akin to the Mercedes-Benz tri-star.

Powered by a 300-horsepower Packard V-8, the Predictor was fully driveable—when it was working right. Ghia had botched the electrical system, so activating any of the fancy servos usually caused a short circuit and great clouds of smoke. Of course, such problems would have been worked out for production, for the Predictor was nothing less than the blueprint for a very ambitious new S-P line.

The Predictor outlined a whole raft of cars built from just three basic platforms: a 130-inch wheelbase for Packard and Packard Executive (the latter basically a detrimmed Patrician), 125 for Clipper and Studebaker President, and 120 inches for Studebaker Commander and Champion. Though hardly original, this plan was a great cost-saver, allowing the three makes to look quite different from each other despite a common inner shell and some shared exterior panels. It was a grand strategy worthy of General Motors—S-P's last attempt to cast itself as a "full-line" producer.

Naturally, the Predictor influence was most evident in the proposed Packards, but the entire line had the show car's general feel. Clipper evolved as something wilder than the Packard, with sharp "shark" fins and more sheetmetal sculpturing in line with Schmidt's aim of appealing to younger buyers in the Dodge-Mercury-Olds class.

The '57 Packard line would have been the same as '56: four-door sedans in Patrician and Executive trim; Caribbean, Four Hundred, and Executive hardtop coupes; and Caribbean convertible. A Patrician/Executive hardtop sedan was slated for 1958, when Executive would divide into Standard and Deluxe series. Also planned was a new factory-built limousine for the "carriage trade" market that Packard had abandoned after 1954; the proposed limo was another Nance ploy to restore the make's once-proud pure-luxury image.

Clipper was first intended to go through '57 as a holdover '56, then be updated for '58. Later, it was decided to offer the new one for '57. In both cases it was planned for just two- and four-door Standard and Deluxe hardtops, though a convertible was sketched and would have been a first for Clipper. Technically, none of these cars would have been Packards, as Nance had registered Clipper as a separate make for '56.

144

Top: *All of Studebaker-Packard's 1957 cars were to have all-new styling, with senior Packards patterned on the '56 Predictor "dream car" devised under Packard styling chief Bill Schmidt. This sketch by staff designer Fred Hudson was one of numerous suggestions for a Predictor-inspired big-Packard face. The* main departure here from the show-car ensemble was the prominent bumper bars with widely spaced "bombs" flanking the slim vertical grille. Above: This undated and unsigned rendering shows an early workout for the facelifted Packard Clipper first planned for '57, which would have *involved another restyle of Packard's existing 1951-vintage "high pockets" bodyshell, the work of John Reinhart. Later on, the game plan was changed so that Clipper would be updated for '57, not 1958, as part of the corporate-wide redesign.*

Down in the low-priced ranks, where a broad lineup was essential for sales, Studebaker was assigned all the above body styles save the limo, plus a station wagon. The pillared and pillarless Hawk "family sports cars," which Raymond Loewy had evolved from his timeless 1953 coupe design, would have returned with few changes from debut-year '56, again in pillared and pillarless styles. There were also plans for a new Express Coupe, reviving Studebaker's light-duty car-based pickup from prewar times. This was set for 1958 as the same sort of station wagon "hatchet job" that Ford used to create its new '57 Ranchero and which Chevy emulated two years later in the El Camino.

In engineering, the '57 Packards would have been advances on 1955-56. Plans called for Bill Allison's effective "Torsion-Level" self-adjusting suspension to be simplified for greater reliability. Ditto the firm's troublesome Twin-Ultramatic transmission. And according to engine designer Bill Graves, Packard's existing 374-cubic-inch V-8 would have been bored to a massive 440 cid for '57, good for at least 300 horsepower and probably far more.

Most intriguing of all was the novel plan for a new V-12, something Packard hadn't offered since 1939. According to Richard Stout, then of Packard Product Planning, this would have been derived from Clipper's smaller 320 ohv V-8, again to conserve scarce cash. As Stout later wrote in *The Packard Cormorant*: "Eight of the cylinders would be bored, then the block moved [halfway down] to do the remaining four. The block was a 90-degree type, 30 degrees off for [the desired] in-step-firing V-12. To compensate, each throw was to be split and staggered 30 degrees

to provide in-step firing. [It was] similar to the principle Buick [later] used to make its existing 90-degree V-6 into an in-step-firing engine." Displacement would have worked out to 480 cid, just seven cubic inches above the 1939 twelve, but with much "squarer" bore/stroke dimensions. Stout added that tooling would have cost only $750,000.

Another bold but ultimately unworkable idea for the big '57 Packards was the "radar brake." This comprised a small grille-mounted radar sensor connected to an electric screwjack that engaged or disengaged the brakes independently of the driver. As Stout related, the radar brake proved itself when a Four Hundred hardtop so equipped was driven at a wall. But later, "a company official drove the [car] home. On his first right turn the sensor picked up a cross-traffic car waiting for a light. Screech! Halt! Recovering, our shaken driver proceeded down a narrow street with parked cars, two-way traffic and pedestrians, all of which alarmed the sensor. . . . Our official was astounded [and] made a beeline for the company garage."

The V-12 wasn't as impractical as the radar brake, but both were forgotten along with the rest of the program when Nance failed to obtain financing. Though not widely known for many years, Nance had proposed selling a "Predictorized" reskin of the all-new '56 Lincoln as a desperate last-minute ploy to attract needed funds. The notion went no further than a single Teague sketch depicting a pretty and remarkably adept blend of two disparate designs.

Packard's own '57 was all but finished long before that trying

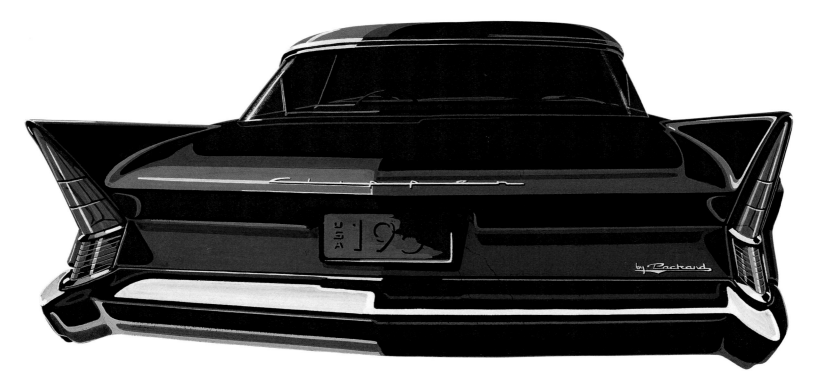

S-P president James Nance had Packard's junior line registered as a separate make for 1956. Clipper was to continue as such for '57, of course, but engineering would have been mostly new and, for the first time, shared with Studebaker's top-line

President series. Staff designer Fred Hudson sketched this finny but clean Clipper stern as one of many renderings he did in 1956 for the ambitious but cost-effective new S-P corporate line. Large "boomerang" vertical taillamps

continued a '56 Clipper motif. The fairly sizeable "by Packard" badge on the lower right of the back panel reflected Nance's '56 strategy of "divorcing" medium-price Clippers from the costlier Packards to restore the latter's high-luxury image.

Top: *An early "theme" sketch for the 1957-58 Clipper looks almost cartoonish, but proposed some new visual signatures for the make, such as flaring "shark" fins and headlamps set high in sculptured fenders. Roofline is a remarkable forecast of the '59 Mercury. Second and third from top: A later, more down-to-earth*

interpretation of ideas in the above rendering, plus a hint of Predictor show car in a squarish roofline with vee'd C-pillars. Front end seems unfortunately blunted, and a showroom '57 Clipper likely wouldn't have looked nearly so long and low as drawn here. Above: Yet

a further evolution of the above concepts, this time signed by Packard Styling and earmarked for Clipper's 1957 Custom series. Clipper would have offered only two- and four-door hardtops in Custom and less-costly Deluxe trim had things gone as S-P planned.

summer of 1956. Again, as engineer Herb Misch later noted, the reason was to give Nance something tangible to show the bankers. What's more, tooling could have been ordered right then, though it's questionable whether Packard could have started production in time for model-year '57.

Among the rubble in the Detroit bunker was a single running prototype of the big new '57. Factory hands dubiously called it "Black Bess." Teague remembered it looking "like it had been made with a cold soldering iron and a ball-peen hammer . . . a last-ditch effort to come up with some money for die models. The doors opened, but it was a very spartan mule. Herb Misch had put it together. There wasn't anything old on it except the V-8. . . ."

The fate of Black Bess is the sort of bittersweet tale that auto-industry insiders find irresistible. Teague said Misch called up

one day in 1956 and told him to see to the car's destruction. But Misch didn't have the heart to carry out the "execution" himself, so Teague called in Red Lux, "an old welder in the studio who had been there since the cornerstone. There were two or three other cars . . . including another black one, a Clipper. I said, 'Okay, it's official, cut the black one up.' I came back around 4 p.m. and he was just finishing. The pieces were lying all around like a bomb had gone off. It was probably the dirtiest trick I ever played, but I said, 'My God, Red, what have you done? Not *this* one, man, the one over in the corner!'. . . . His face drained, and when I told him I was just kidding, he chased me around the room. You've got to have a sense of humor in this business."

Teague always did. He certainly needed it during Packard's last days in the bunker.

Top: *This design sketch from January 1956 shows one proposal for a '57 Clipper convertible. It's graceful, despite prominent fins and other period styling cues. Note the dramatically wrapped windshield. Above left and right: The Packard Predictor show car was the ultimate expression of the corporation's styling ideas as the end drew near. The Predictor was an attention-getter at the*

Chicago Auto Show in February 1956— and no wonder, for it boasted a mind-boggling array of innovations and predictive styling touches. When a door was opened, metal panels above the door retracted into the roof to aid entry and exit. The driver-side seat swiveled outward for additional convenience, which doubtless would have been appreciated by women in that era of

billowy skirts. The reverse-slant backlight (later adopted by Lincoln, Continental, and Mercury) was power-retractable. Wide rear roof pillars almost certainly would have hindered aft visibility, but did provide designers Bill Schmidt and Richard Teague with a place for an assertive new Packard logo. Tailfins were nothing if not aggressive. Inside, pushbuttons dominated a chromed dash.

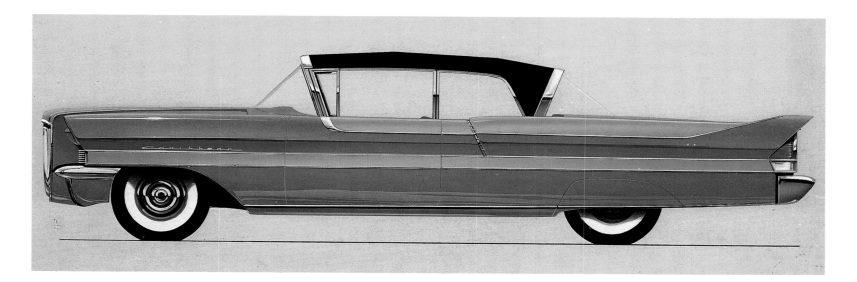

Packard's largest and costliest 1957-58 models were naturally planned to have the strongest visual links with the handsome Predictor show car, as seen in these sketches for a new-generation Caribbean convertible (top) and Four Hundred hardtop coupe (center). The second hardtop rendering (above) is labeled "Proposal A, '58 Packard," and projects only the minor trim changes that are usual for a new design in its second season. Note the variation on the "cathedral" taillamp design first used on the '55 Packards as heavily facelifted by Dick Teague. "Circle-V" emblem on the hardtop C-posts was Teague's attempt at a new "timeless" automotive symbol like the famous Mercedes-Benz tri-star.

1931 Peerless V-16: More Than the Name Implied

Built in the winter of 1930-31, just before Peerless decided to leave the car business, the prototype V-16 was perhaps inspired by the high early acclaim won by Cadillac's new 1930 Sixteen. Production was eyed for as early as 1932 at a reported base price of $3800, which would have made the Peerless quite a bit cheaper than comparable V-16 Cadillacs. Alas, the decimated Depression-era luxury market left no hope for anything like profitable sales volume for this beautifully styled and crafted motorcar.

Top: *A grand 145-inch wheelbase allowed the prototype Peerless V-16 to wear imposing proportions and flowing lines. Body design was done by Franklin Hershey, then of the Murphy Body Company in Pasadena, California, who would go on to work at Pontiac in the* Thirties, Cadillac in the Forties, and Ford in the Fifties. Despite its size, the four-door prototype weighed a svelte 4000 pounds. Above: Except for its iron cylinder heads, the Peerless V-16 engine was made mostly of aluminum, as were the car's body and chassis. Other features *included a narrow 45-degree bank angle, forged-aluminum connecting rods, pushrod-operated overhead valves, and nearly "square" bore/stroke dimensions. Displacement was a sizeable 464 cubic inches, rated horsepower 170 at 3600 rpm.*

Top: *Though classically four-square in shape, the prototype V-16 Peerless shows traces of the "transitional streamlining" trend that began holding sway in the auto industry around 1932-33. Traditional fabric roof insert is visible here, but designer Hershey created a predictive and practical touch by cutting* *the door openings up into the roof to ease entry/exit—likely the first occurrence of this now-commonplace feature. Rain gutters were concealed for aesthetic reasons.* **Above:** *More traditionalism was evident in the dash design and driving position, as well as in the top-quality wood, broadcloth upholstery, and* *other materials. Gauges, which lived behind a glass pane, included a speedometer calibrated to 120 mph and tachometer reading to 5000 rpm. Bottom-mount wipers broke with convention to sweep more windshield area than top-mounted blades.*

Top left: *All doors were rear-hinged. Towering 70.5-inch overall height and Hershey's wrapover doors gave easy access to the posh and roomy five-passenger cabin.* **Top right:** *Upholstery was reportedly treated to resist wear and* fading. *Doors were claimed to strengthen the body when closed, thanks to special-design latches.* **Above:** *Underslung final drive and rear leaf springs made for a long, low look despite 18-inch-diameter wheels (aluminum, of course). Solid axles* appeared *front and rear. Massive chassis side rails measured up to 10.5 inches deep. The one-of-a-kind Peerless V-16 prototype is currently on display at the Crawford Auto-Aviation Museum in Cleveland.*

Shattered Dreams: Pontiac's Plans for Future Fieros

Plastic outer panels made Fiero styling easy and inexpensive to change, but the "driveable space frame" made additional body styles rather more difficult and costly than conventional construction. Even so, Pontiac devised a racy Fiero-based speedster in mid-1983, then updated its styling a year later. The handsome red job seen here was built around 1986, with that year's GT coupe lower-body styling, to investigate the feasibility of a convertible Fiero. Besides a fully engineered soft top with modest "stack height," this prototype boasted a left-hinged rear deck that swung up sideways for engine access. But cost factors made production unlikely even if Fiero had continued past 1988.

Top: Pontiac stretched a standard '85 Fiero coupe into this 2+2 as another trial in adapting the car's space-frame to other body styles. Rear side window wrapped up and over to make a moonroof for those in the cramped back bucket seats.

Protruding side scoop never made it to showroom Fieros, but would have been interesting. Center and above: When the axe fell on Fiero in mid-1988, this full-size mockup was one of several designs in contention for 1990.

Highlights included a "softer" nose, elevated rear deck, longer roof with bulkier B-posts, and a revised mid-body character line that fully encircled the car and tied in more neatly with the taillamps.

Top and center: *This fairly extensive facelift of Fiero's 1987-88 styling may have been aimed at '89, certainly 1990. Bespoilered tail with "tunneled" backlight shows a hint of Trans Am (or maybe '68 Dodge Charger), while Ferrari 308/328 overtones appear in the bodyside* indents, whose shape repeats around the front side-marker and parking lamps. Above: *A variation on this basic theme proposed an even closer Trans Am tie-in via a full-width taillamp lens, long a Pontiac hallmark. The visual kinship with the T/A was no doubt intended to* further Fiero's claim as a unique performance car following the major chassis changes of 1988, rather than its early image as a low-cost collection of corporate components in sporty mid-engine format.

Also in the running for 1989 or '90 was this version of the "Trans Am look" that was being evolved when Fiero was felled by GM cost-cutters. Flared rocker bottoms furthered the T/A connection (top and above), but a more sculptured back panel with triple-lamp motif (middle right) would have been Fiero's own. Front end (middle left) was a careful evolution of 1987-88. Headlamps would have remained hidden beneath pop-up hood panels, but the switch from squares to oblongs shown here implies a planned change to smaller, newer-design high-intensity lighting. All four of these views were taken just outside the Design Staff offices at GM's Warren, Michigan, Technical Center.

Studebaker's Last Stands: Too Late To Save the Day

This chapter concerns what were once called the "cars that could have saved Studebaker." In truth, they could not. By the time the first of these projected designs was broached in late 1963, Studebaker's auto business was virtually beyond saving by any means.

Its problems proved fatal soon enough: fast-falling demand for an aging group of cars, insufficient cash for getting out more competitive new designs, scant hope of finding a funding angel, and a general loss of public confidence—the classic corporate death knell. With that, Studebaker was forced to close its century-old South Bend, Indiana, factory in December 1963 and retreat to Hamilton, Ontario, Canada, where it hoped to make a stand building "common sense" family compacts at the rate of about 20,000 a year. But when it failed to achieve even that modest goal, the firm had little choice but to quit the car business for good, which it did after model-year '66. In the end, Studebaker succumbed to the accumulated effects of some 40 years of inept management, steadily weakened by battling one debilitating disaster after another.

Ironically, the designs shown here might well have turned things around had they reached production under more astute management. They were certainly imaginative cars, and most were exactly right for the times. Even more ironic, Studebaker had finally gotten such astute leadership when the dynamic Sherwood H. Egbert was named company president in 1961 after holding the same job at Los Angeles-based McCulloch Corporation, which sold superchargers to Studebaker. For a brief moment, the situation seemed hopeful.

Egbert was an old friend of noted Milwaukee-based industrial designer Brooks Stevens, and he promptly called on Stevens to remodel the company wares for 1962 as a first step in remodeling Studebaker's sagging image. It was a crash program, less than a year long, but Stevens came through brilliantly. First he created the handsome Gran Turismo Hawk, using the old Raymond Loewy hardtop design from 1953. He also deftly reskinned the compact Lark, then further improved its looks for 1963-64.

But this was only a prelude to Stevens's plans for 1964-66: "As soon as Sherwood Egbert called me in to facelift the '62s, our studios got to work on projections for all-new, or at least new-looking, Studebakers. . . . By degrees, each more radical than its predecessor, these cars would have replaced the Lark, falling at the Cruiser end, the big end of the intermediates. Wheelbase would have been 116 inches, later adding [a 113-inch chassis]. We planned to continue the 289 V-8; though it was old, it was a good engine, and with a blower it went like hell. We . . . mounted [it] farther back for better weight distribution, and prepared three prototypes: a wagon, sedan, and hardtop coupe. Each model had two different sides representing standard and deluxe versions."

Roundly approving what Stevens had in mind, Egbert asked him to oversee construction of full-size mockups. Trouble was,

Studebaker was then so short of cash that all it could afford was $50,000 for the lot. Stevens decided his only hope was Italy, but not some high-priced outfit like Pininfarina. Happily, he discovered a small coachworks in Turin called Sibona-Bassano. "I walked in," he remembered, "and there was laundry on the line and chickens running around. I took these two little guys out and fixed them up with Camparis. We got good prices out of them— $16,500 per car, an incredibly low figure." Better yet, the finished models were worthy of a Pininfarina. Stevens termed them "jewel-like," and recalled Egbert being very excited about them.

As planned, the least radical of these proposals was the '64 design, which was done as a station wagon with sliding rear roof panel *a la* Stevens's new-for-'63 Lark Wagonaire. Grillework continued the "Mercedes look" from his '62-model Lark facelift, but in a more exaggerated trapezoid tapered in toward the bottom. A broad chrome grille header bearing the Studebaker name was spread over to crown side-by-side quad headlamps. Hood and deck were broader, flatter, and lower than on late Larks, while front fenders were sharper and thrust rakishly forward at the top. Mindful of Studebaker's threadbare budget, Stevens contrived to save money by using identical bumpers at each end and center-hinged doors that interchanged diagonally (right front to left rear, left front to right rear).

Opening those doors revealed a modest evolution of the '63 Lark interior, which Stevens blessed with a nifty oblong gauge cluster containing round dials and rocker-switch minor controls. The '64 proposal retained these items, but with gauges grouped in a three-element panel, as on the GT Hawk, instead of a flat one, with outer ends again angled in slightly to enhance legibility. With the doors opened, the area around the gauges lit up as extra courtesy lighting. Those doors were quite thin, contributing to passenger space that was relatively colossal for a compact package. Equally generous glass areas added to the spacious feel inside and made for panoramic viewing to the outside.

Had all gone as planned, this design would have been replaced for 1965 by a slightly more advanced version. Stevens modeled it as a hardtop sedan with broad rear roof quarters, as on the GT Hawk and Ford Thunderbird. An ultra-low beltline and glassy greenhouse were again on hand. So were diagonally interchangeable center-opening doors (complete with vent panes), but here they were cut into the roof for easier entry/exit. Equally predictive were hood and trunklid "cuts" that included the tops of the fenders, giving big openings and easy access to engine and luggage. Up front was a narrower but still large grille of roughly squarish shape, filled with a mesh-and-bar latticework made convex at the horizontal centerline. Outboard were French Cibié rectangular headlights, though such things were then illegal in the U.S.

Predictably, the '65 interior also took proposed '64 concepts a

In 1963, Brooks Stevens, the renowned Milwaukee-based industrial designer who'd long consulted with various independent automakers, came up with a three-phase plan for the 1964-66 Studebakers at the behest of company president Sherwood Egbert, who was looking for a way to revive South Bend's flagging fortunes. Stevens also had prototypes built for pennies, which was about all the firm could afford by then. Step one was this clean, trim wagon projecting the '64 line. Though clearly related to the Stevens-facelifted 1962-63 Lark compacts, bodywork was all-new, with a lower beltline, more glass, and cost-saving diagonal-interchange doors. Novel sliding rear roof panel was a Stevens idea continued from the '63 Lark Wagonaire.

step further. The driver again faced a large upright nacelle holding rocker switches and straightforward white-on-black gauges (a full set save tachometer), plus a couple of hefty levers. The rest of the dash was a slim, low-set padded shelf. Concealed within was a slide-out "vanity," a drawer-type glovebox divided into big and bigger sections. Each part had its own lid, and the larger one lifted to reveal a makeup mirror. Stevens had first used these ideas on the '63 Lark. More novel yet were the radio and clock, which lived atop the dash in clear semi-spheres. Of course, these items would have been optional, and that was the beauty of this design: no unsightly dashboard "blanks" if you didn't order them. Radio operation was clever: Push down on the bubble for on/off and volume, turn it to change stations. The clock bubble also rotated, allowing everyone to tell the time with equal ease. A final touch was a tilt-adjustable steering wheel, an uncommon feature at the time.

Stevens hoped to usher in a whole new generation of Studebakers with his prototype for 1966, a shapely notchback two-door hardtop that he called Sceptre. Eyed at one point as a replacement for the GT Hawk, it looked smoother and lower than the 1964-65 designs, but was still fairly angular and just as glassy on the same relatively compact platform.

But the Sceptre had plenty of startling departures. They began up front, where Stevens replaced conventional sealed-beam headlamps with a single full-width tube, developed by Sylvania, that gave more light with virtually no glare. (It now seems an amazingly accurate forecast of the "light bars" used on certain Mercurys, though those are just for nighttime "identity," not illuminating the way.) Marking the tail was a similar tube behind a recessed full-width red lens. Equally forward-thinking was Stevens's use of blue polarized-glass insets for the C-pillars, which were transparent from the inside but appeared opaque

from the outside. (You can think of this as a low-tech precursor to the light-sensitive LCD "glass" that appeared 25 years later for various show cars and a few aftermarket moonroofs.) Other distinctions included pointy front fenders, front cornering lights, a wheel-cover motif remarkably like that of Mazda's later rotary-engine symbol, and a big tri-color emblem mounted on a raised circle smack in the middle of the hood.

Sceptre's interior was no less striking. Driving necessities were squarely ahead of the wheel once more, but Stevens now put all the secondary gauges in bubbles, each tilt-adjustable for best viewing. The speedometer was a wide strip-type device on a short stalk that pivoted up from between the two middle bubbles; the stalk could be folded flush with the panel, leaving the speedo just above the minor dials, or raised to put it close to driver eye level—a kind of early "head up display." To the right was a floor console angled in for optimum control reach. Here too, Stevens penned a very clean overall dash with lots of padding to conceal another sliding vanity, but also a "rally table" on the right side. Seats were four vinyl-covered buckets with center sections trimmed in chrome-like Mylar plastic.

With all this, the Sceptre would have been a sensation in 1966. But, of course, it had no more chance of reaching showrooms than Stevens's other "comeback" Studebakers. As the designer recalled, money ran very low sometime in 1963, "and we were suddenly told we'd just have to reskin the Lark again. . . . Of course, you never dared stop. So we kept going on the prototypes even then." But not for long, as the book closed on all these ideas when Studebaker closed its historic South Bend plant.

Stevens wasn't the only one working on new Studebakers. Also in the running were two prototypes for an entire line of 1965-66 models patterned on the glamorous Avanti coupe. These were built under the redoubtable Raymond Loewy, the guiding light

Step two in Stevens's comeback plan for Studebaker took the form of this hardtop sedan displaying design projections for 1965. Though visually similar to the '64- targeted wagon, it was slightly more ambitious in styling and features. Hood and trunklid were cut down into the bodysides to provide larger openings for easier engine and luggage access. Note the sharply creased wrapped rear window and clean chrome-edged fenderlines.

More views of Brooks Stevens's '65 Studebaker prototype, which survives with its companions at the Brooks Stevens Museum in Mequon, Wisconsin, near Milwaukee. Front end (top) has a strong Mercedes air, reflecting Studebaker's late-Fifties/early-Sixties role as U.S. Mercedes importer. Rectangular headlamps were still far in the legal future when this prototype was built in 1962-63. As on the wagon proposed for '64, center-opening doors (above left) could be swapped diagonally, and afforded easy access to uncommon passenger room for a 116-inch-wheelbase compact. Clean, modern dash (above right) featured dual "vanity" gloveboxes, another Stevens idea first seen on Studebaker's '63 Lark, plus a big, bold gauge cluster.

for that show-stopping GT, whom Egbert still retained as a consultant. A talented Loewy team had fast conjured the fiberglass-bodied Avanti while Stevens was facelifting the Lark and Hawk—the second of Egbert's two-step plan to spruce up Studebaker's image. But though Loewy delivered exactly what Egbert wanted, the Avanti itself did not.

Like Stevens, Loewy had definite ideas about new Studebakers, and went overseas to get prototypes built. But where Stevens chose an obscure Italian coachbuilder, Loewy chose the obscure French house of Pichon-Parat, near Paris. The results were full-size notchback and fastback designs configured as a two-door on one side, a four-door on the other.

Loewy tried hard to sell Egbert and other Studebaker executives on these Avanti-styled sedans, and the mockups he showed were practical yet different. Either would have been a logical follow-up to the sensational coupe: clean and handsome, with the same unmistakable design signature. Pressing harder than usual, Loewy trimmed both prototype interiors in full leather, though lack of time and money forced the use of decals in place of dashboard gauges.

The one problem with this idea—aside from Studebaker being almost broke—was the car that inspired it. Though first shown in 1962, the Avanti wasn't genuinely available until 1963, by which time events had rendered it a commercial failure. Production delays weren't entirely to blame. As former body engineer Otto Klausmeyer said later: "The fastback prototype should have been built [first] instead of the Avanti. . . . Yet the Avanti was in production, and had been abundantly rejected by the public before the sedan prototypes were finished. . . . The directors would not approve the sedans because they feared the Avanti influence would be the kiss of death, not because they were a bunch of provincial sod-busters, as most articles about

these cars imply." So in the end, there was no way Loewy could win this particular battle. Like Custer at Little Big Horn, the odds were overwhelmingly against him.

But the persistent Brooks Stevens made one last try. Determined to save the South Bend plant and its workforce, he met hurriedly in early 1964 with Charles "Cast-Iron Charlie" Sorensen, the renowned Ford production whiz of the Thirties who served briefly as president of Willys in the mid-Forties. Together they hatched a revolutionary new small car with a very spacious interior on a 113-inch wheelbase. Introduction was projected for 1967-69.

Variously called "Familia" and just "Studebaker," this boxy but attractively clean-lined sedan was conceived around a plethora of interchangeable parts: hood/trunk, doors, bumpers, headlight/taillight housings, windshield/backlight, even side windows. Sorensen conjured a simple production line to suit the unitized fiberglass construction. Specifically, he devised a carrier that moved four half-body molds around to individual stations for gel coating, outside and reinforcement matting, and bake-oven curing. The carrier turned and positioned the bodies at each step, and also returned them to the starting point. With this plus all the dual-duty parts, a simple proprietary engine and minimum frills, Stevens and Sorensen pegged unit production cost at an ultra-low $585, which meant a probable profit at a retail price of $1100 or less.

"I had tremendous hope for this idea," Stevens said later. "I took the project to the board at the end of February 1964, and they were quite interested. Unfortunately, the financial backers had just breathed a sigh of relief after dumping automobiles at last. There was no way any money was going to be made available for anything on wheels. I quit in disgust. I guess it was too late. It was certainly the wrong time to try."

The third and most radical of the Stevens-designed "comeback" Studebakers was this hardtop coupe that the designer called "Sceptre." Briefly eyed at one point to replace the GT

Hawk, it showed the general shape Stevens wanted for the 1966 Studeys. Novelties abounded, such as a full-width front "light bar" instead of conventional headlamps, a Sylvania innovation that

probably would have had a tough time getting past federal rulemakers had a production version come off. The left side of this prototype was modeled with "deluxe" trim.

Top: *The other side of Sceptre displays simpler "standard" bodyside trim, though this may have been just an alternative to the left-side treatment. C-pillars on both sides carry blue-polarized glass insets, which conferred a more "formal" look without hampering occupant visibility.* Center left: *"Bubble" minor gauges and speedometer-on-a-stalk were just two of Sceptre's novel interior ideas.* Center right: *Sceptre's wrapped-over trunklid conceals sizeable luggage space.* Above: *Vying with the Stevens cars for 1965-66 were these prototypes for "Avanti-styled" notchback and fastback Studebaker sedans created under Raymond Loewy, South Bend's other veteran design consultant. Modeled as a two-door on one side (above left) and four-door on the other (above right), these mockups were left to rot in the South Bend attic after Studebaker left Indiana and, finally, the car business. This was their sad-looking state in late 1984.*

Studebaker for '57: Styling That Stayed in the Studios

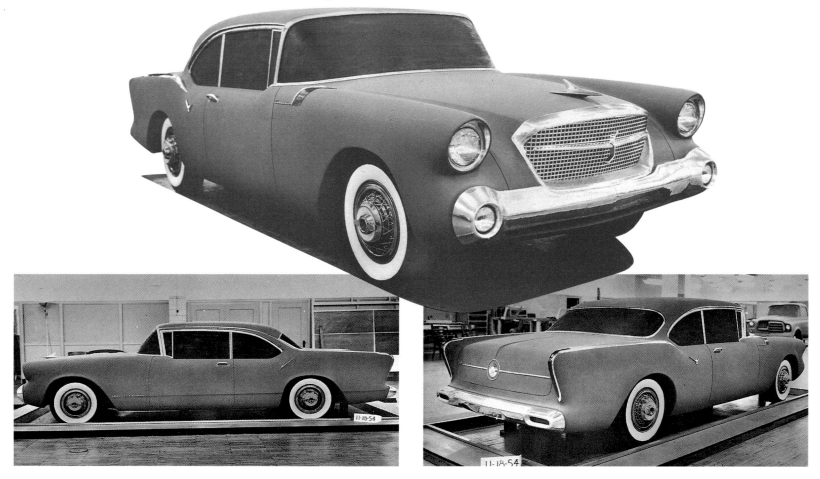

Studebaker employed in-house designers as well as outside styling consultants in the Fifties. One of the latter, the famed Raymond Loewy group, which had been associated with Studebaker since the '39 Champion, proposed this full-size coupe mockup for 1957. Photographed in November '54, it looks somewhat like a '55 Chevy in profile (above left) and faintly Chrysler-like from behind (above right), but step-down beltline and thrusting mesh-filled square grille were unique. Atypical for a Loewy design, management rejected this one, but the front-end theme—minus the "dumbbell" bumper—was approved for the '56 Hawk "family sports cars."

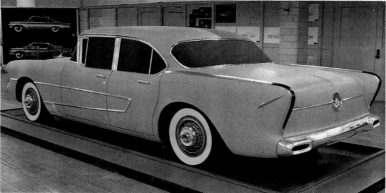

Top row: *The late-'54 Loewy coupe was also planned as a four-door sedan on Studebaker's longer wheelbase. Note the changes to grille (left) and side trim (right). Body contours again have a definite '54-'55 GM flavor. Second row from top and main photo: Shaped under Studebaker chief stylist Duncan McRae, this clay has the basic face adopted for '57 Studeys, plus '58-style* quad headlamps. *However, its roof and windshield shapes suggest a new bodyshell was eyed after plans for a shared 1957-58 Studebaker-Packard platform were abandoned for lack of funds. Rear aspect (except for taillamps) evokes thoughts of '57 Plymouths. Above row: Studebaker planned to revive its Express Coupe car/pickup in the ambitious all-new S-P* line first envisioned for '57. Ted Pietsch *penned five possible accessories for it, including a built-in tilt-up bed cover (left), bolt-on "covered wagon" canvas top (center) and pop-up pup tent (right). Also proposed were a double-door "clamshell" bed cover and a bolt-on wagon-style "camper top" with side windows and glass liftgate.*

Studebaker's Last Truck: One for Westinghouse

Studebaker abandoned more than two sporting automobiles when it fled to Canada in late 1963. It also gave up on trucks. That was sad, for Studebaker had often done well with commercial vehicles, which it began offering a few years after getting into horseless carriages with electric cars in 1902.

Though trucks were always a sideline to South Bend's car business, it might have been the other way around. As marque expert Fred K. Fox observed in *Studebaker: The Complete Story,* the company's "biggest mistake, as far as truck sales are concerned, was made in the Teens, when they did not push boldly into the truck market before they quit the horsedrawn wagon business. Had they done this, they might have come out of the Depression as a major truck producer that built cars as a sideline." As it was, "the health of the truck division always depended on the health of Studebaker's car division."

And in 1963, Studebaker's car division was anything but healthy. The firm gushed some $17 million in red ink that year, its fourth biggest annual loss in the postwar period after disastrous 1954-56. The truck line reflected decades of corporate ups-and-downs by having seen little basic change since 1949, mainly because sales were seldom high enough to justify the costs. Though Studebaker did manage the cute little Champ pickup in 1960, with cab styling borrowed from its year-old Lark compact car, this was just more old stuff in a new wrapper that couldn't hope to generate the needed sales or income. Overall, the company wrote its 1960 ledger in black, but the bottom line was a slim $709,000.

In 1961-62, however, Studebaker rebounded to earn over $5 million, thanks to the energetic efforts of its new president, the hard-driving Sherwood Egbert. As noted elsewhere in this book, Egbert came aboard in 1961 and immediately put the rush on facelifts for Studebaker's existing cars. The eminent designer Brooks Stevens came through beautifully for '62 with a remodeled Lark and a more fully overhauled "family sports car," the retitled Gran Turismo Hawk. Egbert also got fast results from the Raymond Loewy group on the sleek new high-performance, image-boosting car he wanted, and within months the world was applauding the singular Avanti.

These projects depleted funds that might have been used to rejuvenate the Studebaker trucks, but Egbert did what he could, adding a flush-fender "Spaceside" Champ and new diesel-powered light- and heavy-duty models in 1961-62. He also went after more military and government business, and even arranged to repurchase a Studebaker truck factory that had been sold to the Curtiss-Wright combine, which still nominally controlled South Bend's affairs. Trouble was, that plant cost a hefty $7.5 million, yet the order for the half-tracks it was supposed to build was cut in half, and remaining production didn't begin to cover the purchase

price. Egbert continued trying to push Studebaker as a "big truck" producer, but with little success.

The company was staring down a very dark tunnel by mid-1963, when a faint glimmer of light appeared. It came in the form of a $9 million contract to build some 4200 small postal delivery vehicles. Unfortunately, Studebaker's role in the "Zip-Van" project involved mere assembly of a collection of its aging truck components within a tall, square body supplied by Met-Pro of Lansdale, Pennsylvania. Of course, this order was welcome—*any* new business was by then—but it was too modest to offset losses from withering sales of Studebaker's mainstay cars. By November, the clock on the South Bend wall read one minute to midnight.

Just when all seemed lost, Westinghouse popped up with a proposal that might help save the day. Approaching Egbert directly, the president of the big electrical and home appliances maker asked whether Studebaker would be interested in designing and building a fleet of compact, purpose-designed delivery vehicles. Egbert couldn't say yes fast enough. The result was the one-of-a-kind prototype you see here—Studebaker's last truck.

According to an article by the aforementioned Mr. Fox in the February 1983 *Turning Wheels,* the official publication of the Studebaker Drivers Club, there were actually two Westinghouse prototypes: the pickup pictured here and a companion van. The pickup was unearthed in late 1982 by SDC member Terry Chase of Nashville, Michigan, who stumbled across it in one of those proverbial old barns (about 70 miles from his home) that enthusiasts just *know* contain rare treasures like this. The van, unfortunately, has yet to resurface.

But fairly complete records of this project also survive, and they are interesting. Predictably, styling was handled by Studebaker's beleaguered design chief Randall Faurot—the last person to hold that title in South Bend. Seeking high simplicity, doubtless with an eye to minimizing costs for both Studebaker and Westinghouse, Faurot sketched a cab-over-engine (COE) design with nary a curved line on it save headlamps and wheel arches. The result was boxy in the extreme, but the simple flat panels were a stamping operator's dream. Dominating the high bluff front was a big windshield canted forward at the top, suggesting that no thought was given to aerodynamics. Then again, this was long before "airflow management" was deemed important for any vehicle. And besides, these were supposed to be short-haul utility rigs, not high-speed interstate flyers. Interestingly, though, Faurot also sketched an upsized version of the Westinghouse design as a diesel semi-tractor trailer.

Faurot had sketched the pickup with a right-side box panel that dropped down to form a load ramp, as on Chevrolet's rear-engine Corvair-based Rampside, which pioneered this feature on

Photographed in the mid-Eighties near Discovery Hall at the Studebaker National Museum in South Bend, Indiana, Studebaker's prototype 1963 Westinghouse pickup was a simply styled cab-over-engine (COE) design with flat, easily stamped body panels, compact dimensions, and an obviously bluff front. The COE format was likely chosen to facilitate side-by-side production of a planned companion cargo van, which was also prototyped but has so far remained undiscovered. The pickup was rescued from a Michigan barn in 1982 by a member of the Studebaker Drivers Club.

its 1961 debut. However, the completed Westinghouse pickup prototype had a conventional box, plus a less prominent grille; Faurot had envisioned a low, full-width eggcrate with the headlamps at its outer ends. The van was penned with conventional dual center-opening load doors behind the right-side cab door, but was otherwise similar to the pickup.

As usual, the capable Gene Hardig, a veteran of many spare-every-expense South Bend products, handled engineering chores for the Westinghouse project. He made liberal use of the parts bin—again—mainly because he had no other choice. Inevitably, Studebaker's familiar 289 V-8 was plunked directly beneath the cab to drive the rear wheels—somewhat extravagantly—through a three-speed automatic transmission with "Power Shift" manual-hold control, just like in the pricey Avanti. Fox has noted that the prototype uses Avanti parking lights and 1961-66 standard-Studey door handles. Its dash is similar to the Champ panel: a simple affair with a small gauge cluster dead-ahead of a big steering wheel, which sits almost horizontally atop a long column poking up from the floor.

Despite its "big-rig" looks, the Westinghouse pickup is quite compact. Fox reports a trim 95-inch wheelbase and 168-inch overall length, while width measures 72 inches, overall height a towering 78. The pickup box is eight feet long, a standard size suggesting that it, too, came off the shelf.

The Westinghouse project apparently prompted Studebaker

to think about a whole line of "forward control" trucks as a mid-Sixties replacement for its elderly Transtar models. Faurot's diesel-tractor proposal suggests this, as does another artifact from Studebaker's last days in South Bend: a ⅜-scale model for an altogether more comely COE pickup. Rescued by one Ralph Schlarb and later purchased by marque authority Asa Hall, this envisioned simple construction like the Westinghouse design, but wore a curved front swept gracefully upward in an unbroken arc from bumper to roof. It looks nice even now; what a shame that nothing further was done with it.

But curved or boxy, car or truck, no future Studebaker had a prayer of production once the board of directors decided to flee South Bend in the face of cash reserves fallen desperately low. Studebaker thus closed its crumbling high-overhead Indiana complex in December 1963 and forgot all about trucks, not to mention the Avanti and GT Hawk. Egbert was duly fired, and a new management team bet what meager funds were left on heavily restyled Larks built at the firm's Canadian operation in Hamilton, Ontario. But it was all in vain, and after 114 years, Studebaker gave up on wheeled vehicles altogether in mid-1966.

Today, the prototype Westinghouse pickup sits cheek-to-jowl with other historic Studeys at the Studebaker National Museum in South Bend. It's a place well worth visiting, if only to be reminded that innovation so often flourishes in adversity—and that no enterprise is immortal.

Randall Faurot, the last head of Studebaker styling in South Bend, penned the Westinghouse pickup with simple lines for easy, low-cost assembly. His original design had the right side of the cargo box, just behind the cab, folding down to form a load ramp, but lack of time and money precluded this idea on the prototype. The project was born when Westinghouse commissioned South Bend for a fleet of purpose-designed short-haul utility trucks and vans, but Studebaker also used it as a springboard toward a possible new line of medium and heavy trucks for the mid-Sixties.

Top: *Though obviously un-aerodynamic with its big bluff front, Studebaker's 1963 prototype pickup afforded great driver vision. Cab was front hinged to tilt forward as a single unit for service access.* Above: *Bus drivers would have felt at home inside (left) and cab step-in was rather narrow (middle), but the agreeably simple Westinghouse pickup promised low running costs, just what commercial truck operators want. Gauges (right) were confined to a modest cluster dead ahead of the wheel. Fully driveable, the prototype pickup carried Studey's old but reliable 289 V-8, mated to a three-speed "Power Shift" automatic.*

Tucker: What Really Happened and Why

The backlot of automotive history is littered with failures. Some deserved their fate, others didn't. Then there's the Tucker. Unveiled in 1948, it bristled with advanced features, some not copied for many years, others not copied at all. To war-weary Americans expecting bold new postwar ventures, it looked every inch like the long-promised "car of the future." But the Tucker had no future at all. Surrounded by controversy that raged for decades, it never really got a chance to succeed or fail.

At the heart of this story is Preston Thomas Tucker, a fast-talking 200-pound six-footer worthy of another P.T.—Barnum. An "autoholic" since boyhood, Tucker clerked at Cadillac Engineering, worked on a Ford assembly line, then sold Studebakers, Packards, and Dodges before working up to sales manager at Pierce-Arrow. Because fast cars were his first love, he was well-known to famous drivers and other denizens at the Indianapolis Speedway. In 1937, after serving as a police officer, Tucker went home to Michigan, settling in Ypsilanti, near Detroit, where he designed a high-speed military scout car. The military didn't buy it, though its novel gun turret was adapted for World War II aircraft.

While America was at war, Tucker was busily forming ideas for a radical new car to be engineered by friend Harry Miller, famed designer of Indy racing engines. When Miller died in 1943, Tucker found a new engineer in Ben Parsons, then forged ahead. In December 1945 he boldly announced plans to produce "The First Completely New Car in Fifty Years," then spent the next 12 months securing a plant, lining up suppliers, hiring talent, and building a prototype long since known as "The Tin Goose."

What emerged as the Tucker "48" was a sleek fastback four-door sedan styled by young Alex Tremulis, late of Briggs Body Company. Riding a rangy 128-inch wheelbase, it stretched 219 inches long overall and stood just 60 inches high—quite low for the day. The sleek missile shape suggested the "Torpedo" name used in early advertising.

That shape was partly owed to a compact rear-mounted drivetrain and all-independent suspension, features evidently inspired by the successful prewar racing Audis and Mercedes of Dr. Ferdinand Porsche. The Tin Goose carried a whopping 589-cubic-inch Miller-designed flat-six with 150 horsepower, but it couldn't be perfected in time. Tucker responded by purchasing a Syracuse, New York, firm, Air-Cooled Motors. This company was a descendant of the Franklin automaking concern, which had turned to engine-making after curtailing auto production following the 1934 model year; Preston acquired the company in order to obtain its 335-cid engine, similar to the Goose's flat six and proven in wartime Bell helicopters. An all-alloy design weighing just 320 pounds, this unit was converted to water-cooling, giving Tucker the industry's first sealed cooling system. Outputs were

166 bhp and a thumping 372 pounds/feet of torque—and that on mild 7:1 compression; squeezes of up to 10.5:1 were possible. The chassis was a sturdy, box-section perimeter type with subframes carrying suspension and drivetrain. The transaxle was a four-speed unit borrowed from the late front-drive Cord 810/812, but a self-shifting "Tuckermatic" found its way into a few of the 50 production Tuckers ultimately built.

Lack of time, money, and even technology precluded some of Preston's more way-out notions, such as separate cycle-type front fenders, curved windshield, disc brakes, collapsible steering column, and "Torsilastic" rubber springs instead of conventional steel coils. Even so, the Tucker was plenty unusual. A central "cyclops-eye" headlight turned with the front wheels, doors were cut into the roof to ease entry/exit, a roomy six-passenger cabin came with a "step-down" floor, and front and rear bench seats interchanged for even upholstery wear. Safety features abounded at Preston's insistence: massive bumpers, recessed pushbuttons instead of interior door handles, control knobs on a drum instrument cluster tucked out of harm's way ahead of the steering wheel, windshield glass that popped out harmlessly on impact, and a "spacious Safety Chamber" where front passengers could dive "in case of impending collision."

Despite its size and 4200 pounds of heft, the Tucker was relatively speedy. The generally accepted performance figures are 0-60 mph in about 10 seconds and top speed of at least 120 mph. The latter was owed in large part to Tremulis's aerodynamic styling with its estimated drag coefficient of just 0.30—good even now. In July 1948, the factory ran eight cars at Indy around the clock for two weeks. They sped up to 115 mph, yet averaged 20 mpg at 50-55 mph and a creditable 17-18 mpg at 80-90 mph. Just as nice, the all-independent suspension, lightweight rear engine, and center-point steering made for surprisingly easy handling and secure roadholding.

If this car was so good, why did things go so wrong? The answer is the way Preston Tucker conducted his affairs. He issued $15 million in stock to lease a huge wartime Dodge plant on Chicago's South Side, then somehow transformed the lease into a purchase with little money changing hands. Though perfectly legal, the transaction raised eyebrows at the Securities and Exchange Commission (SEC), which was on the lookout for fast-buck artists in the booming postwar economy. Meantime, Preston raced to start production—which finally began in March 1948—only to be stymied when the War Assets Administration (WAA) refused his bids on a surplus blast furnace and an idle steel plant. Then, in June, network radio commentator Drew Pearson reported a tip that the Justice Department was going to "blow the Chicago auto firm higher than a kite." Less than a week later, the SEC began an investigation, charging Tucker with "fast-sell" tac-

Once owned by San Francisco's Chart House Museum and last known in the hands of a private collector, this is number 41 of the 50 production Tuckers built—and one of a handful that were hand-assembled after the firm was forced to close its Chicago plant in August 1948. Styling by Alex Tremulis bristled with advanced features, including doors cut into the roof, "cyclops eye" center headlamp, and a slippery torpedo shape whose estimated drag coefficient (0.30) is respectable even today.

tics. This and a mountain of unwanted publicity soon forced Tucker to close up after only 37 "production" cars went down its very short assembly line.

In the aftermath of the SEC probe, Tucker and seven associates went to trial in October 1949 on 31 counts of conspiracy and securities and mail fraud. History suggests the inquiry was abetted by a friend of SEC commissioner Harry McDonald: Michigan's Senator Homer Ferguson, who perhaps feared that the Tucker would live up to its promise, be a huge success, and cost his Detroit constituents a lot of sales and jobs.

Regardless, the jury declared the trial a farce, and fully acquitted Tucker and his colleagues on a windswept Chicago day

in January 1950. Ironically, Tucker Corporation still had funds to produce its car, and receivers briefly considered going ahead. But public confidence was long gone, so everything was auctioned off at 18 cents on the dollar.

Tucker's failure contrasts greatly with the early success of that other postwar upstart, Kaiser-Frazer, which united veteran auto man Joseph Frazer and billionaire construction magnate Henry Kaiser. For one thing, K-F began with much deeper pockets than Tucker, with initial capitalization at $52 million as opposed to Preston's $15 million. And even that wasn't enough to properly launch K-F; Henry later said he should have started with $200 million. Still, K-F was always the far better credit risk in the

Initially called "Torpedo," the Tucker was unveiled to startled gasps and great applause in early 1948—abetted by plenty of hoopla from fast-talking, promotion-wise Preston Tucker. The "Tin Goose" prototype shown here

differed in numerous ways from production models, but the public was briefly convinced that Preston had indeed delivered "The First Completely New Car in Fifty Years." Though women were

long used gratuitously and chauvinistically in such publicity events, Tucker employed one of the first women to hold a major post in the auto industry, interior designer Audrey Hodges.

Top: *A look at Tucker number 41 from behind. Prominent grille served notice of the engine's location and helped exhaust heat from the rear-mounted radiator in America's first fully sealed cooling* *system. Intake air ducted from the largish grilles at the front of the rear fenders.* Above: *These close-ups highlight the Tucker's front "Safety Chamber" and tidy driving controls (left), spacious aft* *cabin* (center), *and big water-cooled, rear-mounted flat-six engine* (right). *Front and rear seats could be swapped so owners could even out upholstery wear.*

government's eyes. Indeed, the WAA awarded K-F the very blast furnace and steel plant it had refused Tucker out of doubt that he could pay for them.

To be sure, Tucker's total resources were always far less than K-F's. For example, Tucker needed two-and-a-half-years to produce its first cars; K-F produced its first ones just 18 months after incorporating. Although Preston promised to hire upwards of 35,000 workers, his maximum workforce never rose above 2000; K-F's approached 20,000. The upshot was that production over the first two full operating years totaled 50 for Tucker versus some 160,000 for K-F.

Most damaging of all was Preston Tucker's refusal (or inability) to conduct his firm's affairs in a professional and forthright manner. The company floated a single stock issue, made many over-optimistic promises, and got caught fiddling with the books. (At the trial, vice-president Herbert Morley testified to some $800,000 in unaccounted-for receipts in 1947 alone.) As if that weren't shady enough, Tucker also made indirect payments to promoters and planned to assign work to his mother's machine shop in Ypsilanti (which couldn't have handled it anyway, another bone of contention with Morley).

K-F, in contrast, frankly admitted that its first two stock issues were risky investments, but clearly listed what it had done already and what it hoped to accomplish in the future.

Preston Tucker lacked the Washington contacts that had been cultivated by shipbuilder Henry Kaiser. Worse, he lacked Joe Frazer's patience and tact, and the sort of practicality that characterized the K-F management team. Apparently oblivious to reality, Preston was still promising 1000 cars a day as late as mid-1948, when his dream was all but dead.

In the end, Preston Tucker was, as *Special Interest Autos* magazine said in 1973, "essentially a small-time promoter . . . out of his

pond. He remained a stranger and perhaps even a threat to the SEC, and he didn't know anyone in government. He was careless in some of his pencilwork [and in] his talk too, and when the SEC jumped on him about [15 initial stock irregularities], those irregularities did exist."

As for the Tucker automobile, history has deemed it a good-faith effort to offer something new and maybe even advance the state of the art. It was brilliant in many ways and lived up to Preston's claims, but whether it would have succeeded in the marketplace remains debatable. Even if Tucker could have delivered it for $2500 as promised ($4000 seems more realistic), the public would surely have resisted something so radically different until it was thoroughly proven.

Though broken by his Chicago ordeal, Preston vowed to build another car. Tremulis penned a stunning fastback coupe called Talisman as a proposed new Tucker for the Fifties but, for obvious reasons, nothing came of it. Preston then moved to Brazil, where he hoped to build a sporty two-seat kit car called the Carioca. Sadly, that and his other plans died when he succumbed to cancer on December 26, 1956. He was only 53.

More than three decades later, director Francis Ford Coppola turned Preston's story into a major motion picture. But though good drama, *Tucker: The Man and His Dream* (released in 1988) is sometimes bad history. Tremulis, for instance, had not already designed Preston's car before he met Tucker; he did so just before New Year's 1947—and in only five days. At least you shouldn't worry about the Tuckers that get smashed in one scene; they're replicas. Those in the concluding "parade" sequence *are* real: most every Tucker built, including two owned by Coppola himself. Not that Preston ever led his precious few cars in triumphant procession the day of his acquittal—let alone in the warm sunshine of January Chicago—but that's Hollywood for you.

Left: *Reporters and invited guests turned out* en masse *for the gala Tucker press preview in early 1948, staged in a downtown Chicago hotel ballroom.*

Right: *In attendance that night were, naturally, Preston Tucker himself (far left), one Art Baker, Preston Jr., and Charles T. Pearson, who served as Tucker*

Corporation press agent. In 1960, Pearson recounted his experiences and the Tucker debacle in a book called The Indomitable Tin Goose.

A final look at number 41. Most every Tucker built starred in the ending scene of Francis Ford Coppola's 1988 docudrama Tucker: The Man and His Dream. *The unique "cyclops eye" moved with the front wheels to light the way through turns. Preston Tucker had also wanted freestanding cycle-type fenders, "Torsilastic" rubber springs, curved windshield, and disc brakes, but those and other notions were ruled out by lack of technology, money, or both. Alex Tremulis penned the essential shape in just five days in late December 1946. The "Tin Goose" prototype was built in a relatively quick 100 days.*

Top row: *Consultant designer "Dutch" Darrin offered these ideas for a 1955-56 Aero-Willys restyle. Sedan mockup (left) combined elliptical front wheel "speedlines" from Kaiser's Henry J compact with the distinctive "rosebud" grille from Dutch's own Kaiser-* *Darrin sports car. Hardtop (right) was remodeled with a Studebakerish tail featuring lengthened fenders and downsloped deck.* Other photos: *A two-sided early scale clay (middle row) and a near-final workout (above) for the* rebodied Brazilian Aero by consultant Brooks Stevens, which saw production as the Willys "2600" beginning in 1963. Note the scale model's 1955 U.S.-style grille, but also the squared-up rear roof and fenders.

Here's the Stevens-restyled Aero-Willys 2600 as introduced to Brazil in 1963. Though the body differed considerably from U.S. models, chassis and running gear were carryovers, albeit modified to suit Brazilian conditions. This car did good business throughout the Sixties as the evolved Itamaraty model, which then became a Ford product when Dearborn acquired Willys do Brasil. Production lasted all the way through 1972, by which time the Aero's basic enginering was 20 years old.

Willys quit the U.S. car market after 1955, but it never quit building Jeeps— or considering new Jeep offshots. When the firm began looking to replace its late-Forties Jeep-based trucks, Brooks Stevens penned a compact cab-over-engine design that emerged in 1958 as the FC-Series,

America's first light-duty COEs. Left stillborn was a passenger version mocked up as a full-size clay by 1952 (top left and right) and likely seen as a replacement for the old Station Sedan. Original four-door style evolved into a six-door configuration by 1956 (other

photos), by which time it had grown a vestigial rear deck. Production FCs had dual headlamps, not quads (center left). Small "110" badge behind driver's door (center right and above) likely denoted wheelbase length.

Though its "Forward Control" trucks inspired early-Sixties designs from Chevy, Dodge, and Ford, Kaiser-Willys (later Kaiser-Jeep) never got around to a passenger wagon that would have been America's first minivan. Still, designer Brooks Stevens pushed hard for the idea with the "Commuter" (top and center).

Evolved from his earlier "110" proposal, it rode a shorter wheelbase but retained six doors, which made the rearmost door openings quite narrow at the bottom owing to intrusion from the back wheel cuts. Side window shapes were also changed and styling details made more

car-like, but Toledo somehow couldn't see the need for a minivan. Stevens also proposed a light-duty Commuter pickup (above), likely to make the project more financially appealing via costs amortized over higher volume, but again the word was "No sale."

After leaving the U.S. car market in 1955, Willys repaired to Brazil to build modified Aero passenger models as well as a restyled version of the Jeep-based Station Sedan wagon called Rural. In the Sixties, Willys do Brasil expressed interest in a new Jeepster derived from the Rural in much the same way that Brooks Stevens had related the original U.S. Jeepster to the Station Sedan. Stevens flew down to Rio and sketched the idea on his hotel room dresser in a single evening. This remarkably life-like scale model shows how it would have looked had management given a final nod. Proportions and general feel are as for the earlier U.S. Jeepster, but the fenders and Rural front end obviously differed, and there was a new, more modern dash. The project was abandoned for want of sufficiently sizeable demand.

Top left and right: *Two more views of the scale model for the aborted mid-Sixties Brazilian Jeepster. Slightly curved one-piece windshield was one of many differences from the original U.S. design. Center and above right: AMC took over Jeep Corporation in 1970, then showed this sporty concept called XJ001. Built on a production CJ chassis with an* 81-inch wheelbase but wider tracks, it carried a 360 AMC V-8 and a rakish fiberglass body with stripes galore but no doors or top. Strictly an "idea" piece, the XJ001 premiered at the 1970 New York Auto Show. Above left: *A distant forecast of the modern Wrangler was this "Concept Jeep II," a late-Seventies AMC exercise built to test the marketing waters* for something smaller, lighter, and thriftier than the then-current CJ. "Back to basics" design with simple body and fold-down windshield echoed the original wartime Jeep, but CJ loyalists didn't see the need for it. Worse, AMC couldn't spare the cash to produce it due to dwindling sales of its mainstay passenger-car lines.

INDEX

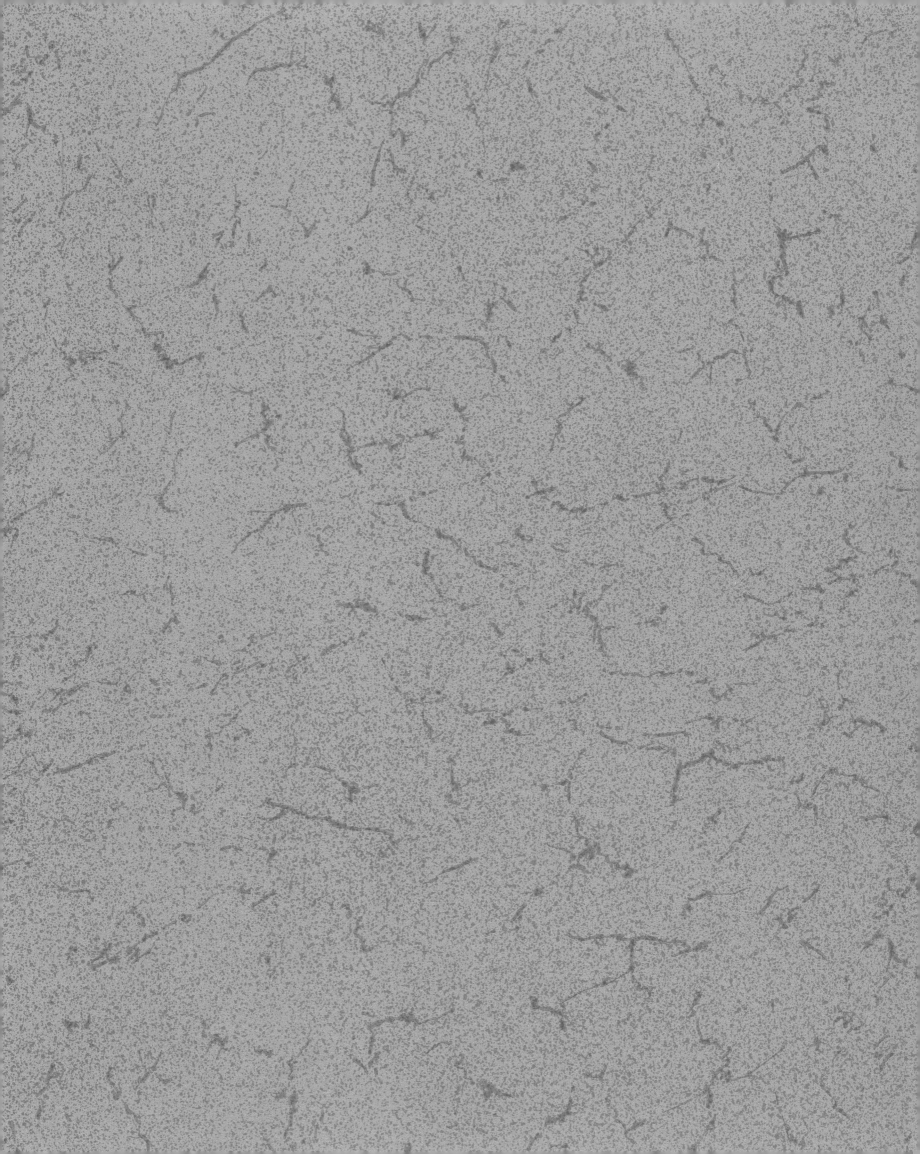